Microdosimetry

Experimental Methods and Applications

Microdosimetry
Experimental Methods and Applications

Lennart Lindborg
Anthony Waker

CRC Press is an imprint of the
Taylor & Francis Group, an **informa** business

CRC Press
Taylor & Francis Group
6000 Broken Sound Parkway NW, Suite 300
Boca Raton, FL 33487-2742

Printed on acid-free paper

International Standard Book Number-13: 978-1-4822-1740-7 (Hardback)

Library of Congress Cataloging-in-Publication Data

Names: Lindborg, Lennart, author. | Waker, Anthony, author.
Title: Microdosimetry : experimental methods and applications / Lennart Lindborg, Anthony Waker.
Description: Boca Raton, FL : CRC Press, Taylor & Francis Group, [2017] | Includes bibliographical references and index.
Identifiers: LCCN 2016058615| ISBN 9781482217407 (hardback ; alk. paper) | ISBN 1482217406 (hardback ; alk. paper) | ISBN 9781315373737 (e-book master) | ISBN 1315373734 (e-book master) | ISBN 9781482217476 (ePub) | ISBN 1482217473 (ePub) | ISBN 9781482217438 (Web PDF) | ISBN 1482217430 (Web PDF) | ISBN 9781315321622 (Mobi/Kindle) | ISBN 1315321629 (Mobi/Kindle)
Subjects: LCSH: Microdosimetry--Textbooks. | Radiation dosimetry--Textbooks.
Classification: LCC QC795.32.R3 L54 2017 | DDC 612/.014486--dc23
LC record available at https://lccn.loc.gov/2016058615

Visit the Taylor & Francis Web site at
http://www.taylorandfrancis.com

and the CRC Press Web site at
http://www.crcpress.com

Contents

Preface

Somewhat paradoxically, microdosimetry can be viewed as both a fundamental and a highly specialised niche field of study within radiation science. For this reason, many of the classic texts in radiation physics and dosimetry discuss microdosimetry only in a cursory manner or not at all and researchers requiring an introduction and reference text have to rely on original sources that, in spite of their excellence, have become increasingly out of date or difficult to obtain. We have both been teaching the principles of microdosimetry for many years and the primary motivation for writing this book was to provide an accessible single-volume account of experimental microdosimetry for use by graduate students and colleagues in the radiation sciences. That said, this book would not have been written if not for a period of collaboration between LL and Prof. Hooshang Nikjoo whose willingness to share his knowledge and insights of theoretical microdosimetry and biophysics was decisive for this work and is sincerely acknowledged.

The impetus to complete this task came from Professor Nikjoo, who recognised that a new generation of students was keen to use microdosimetry as an investigative tool in their radiation protection, radiation biology and radiotherapy research and arranged for an international graduate course to be run at the Karolinska Institutet in June, 2012. The course covered theoretical and computational facets of microdosimetry as well as laboratory exercises introducing different experimental techniques. It was these basic laboratory exercises that convinced us that a text dedicated to experimental aspects of microdosimetry alone could serve as a valuable resource for both new and established researchers. It is furthermore our hope that researchers involved in theoretical and computational microdosimetry also gain some benefit from this text as they acquire knowledge concerning the various uncertainties associated with measured values of microdosimetric quantities and to which they often compare or benchmark their own results.

To meet these objectives, we have structured the book to cover first the basic quantities used in microdosimetry and their relationship to quantities used more generally in dosimetry and radiation metrology. This is followed by a description and discussion of the instruments, methods and procedures used as well as the latest developments in detector technology. The second half of the book covers applications in radiation biology, radiation therapy and radiation protection.

Both of us have been extraordinarily fortunate in having been able to carry out experimental microdosimetry research throughout our scientific careers and early on to have come under the tutelage of some of the founding figures in the field. We hope in some small way we have been able to communicate the fascination and deep satisfaction, as well as the value, of this branch of radiation science that we ourselves have experienced. To complete this work, we have had to rely on the good will of many of our long-standing colleagues and collaborators and former students, all of whom graciously gave of their time to review and comment on what we had written. We would especially like to thank Steve Marino (Columbia University), Pascal Pihet (Institut de Radioprotection et de Sûreté Nucléaire [IRSN]), Hans Menzel (International Commission on Radiation Units & Measurements [ICRU]), Graeme Taylor (National Physical Laboratory [NPL]), Jan Lillhök and Jan-Erik Grindborg (Swedish Radiation Safety Authority [SSM]); their collective knowledge and insight has greatly improved our original efforts and it goes without saying that any deficiencies or errors that remain are entirely our own. We would like to express our sincere thanks to the staff at Taylor & Francis (Francesca McGowan, Rebecca Davies and Emily Wells) without whose constant encouragement, firm but gentle prodding and unfaltering cheerfulness the publication of this work would never have been achieved. Similarly, we wish to recognize and thank Adel Rosario for carrying out the exacting task of copy-editing the final manuscript with precision and amazing patience.

One of us (AJW) acknowledges the generous support and encouragement for scholarship and research given by the University of Ontario Institute of Technology (UOIT); the Natural Sciences and Engineering Research Council of Canada (NSERC); and the University Network of Excellence in Nuclear Engineering (UNENE) and one of us (LL) acknowledges with gratitude the library staff of the Karolinska Institutet for their bibliographical help and support given so open-heartedly. Finally, although words alone are quite insufficient, we would both wish to express our deep debt of gratitude to our respective spouses and life-partners, Anna-Lena and Debra, for their patience, encouragement, and understanding over many decades of our single-minded quest for a fuller understanding of radiation physics and biology.

Lennart Lindborg
Stockholm, Sweden
Anthony Waker
Oshawa, Canada

CHAPTER 1

Introduction

Microdosimetry as an experimental branch of radiation science arose more than half a century ago from efforts to measure the quantity Linear Energy Transfer (LET). At that time, LET was becoming recognised as the best way to characterise and quantify 'ionisation density' or 'radiation quality' as well as a means by which to understand the physical basis of why one radiation type may be different from another in its biological impact. Harald Rossi, considered the founding father of microdosimetry, along with other pioneers in the field, quickly realised that the experimental methods they were developing provided a distinctive insight into the stochastic nature of radiation interaction as well as a unique signature of radiation quantity. Albrecht Kellerer, a principal figure in formulating the theoretical foundations of microdosimetry and a longtime collaborator of Rossi's, has given an insightful overview of the origins and evolution of microdosimetry in the first Rossi Lecture delivered at the 13th Symposium on Microdosimetry (Kellerer, 2002).

In the half-century or so since the early days of microdosimetry, the field has devolved into two main distinct lines of inquiry. Rossi and Zaider, in their text *Microdosimetry and Its Applications* (Rossi and Zaider, 1996), use the terms *regional microdosimetry* and *structural microdosimetry* to describe these branches of research activity. Regional microdosimetry is concerned with the measurement and study of the stochastic energy deposition in sites of microscopic size and is principally an experimental measurement science. The measurement methods developed have been applied in a number of areas of radiation research, and because this activity stems directly from the early days of microdosimetry it is often referred to as 'classical microdosimetry'. This book will deal almost exclusively with the techniques, methods and applications of classical microdosimetry. Structural microdosimetry is concerned with the detailed pattern of radiation interaction and energy absorption in matter and the overlay of this pattern with the detailed molecular structure of the target itself. Consequently, structural microdosimetry is much more readily studied through theoretical methods such as track-structure modelling and biological effect modelling using Monte Carlo techniques. However, even advanced methods of computational modelling require some experimental validation, and the measurement methods developed in classical microdosimetry are still used to advantage.

1.1 EXPERIMENTAL MICRODOSIMETRY AND RADIATION QUALITY

As inferred in the preceding text, experimental microdosimetry is inextricably linked with the concept that the biological effects of radiation are the result of energy deposition in sensitive sites or *targets* of specific size. Indeed, a constant theme in microdosimetric research has been the search to identify the size of target that can be most readily correlated with observed biological effects. One conclusion of this research activity is that there is probably no single target size that can be associated with all biological effects; however, the scholarly effort involved has been a principal driver behind many of the innovative experimental studies in radiation biology that have informed us of the early physical events that determine the final biological outcome of exposure to ionising radiation. An excellent account of the study of radiation effects and target size over the past 50 years can be found in the 2nd Rossi Lecture given by Dudley Goodhead at the 14th Symposium on Microdosimetry (Goodhead, 2006). Another result of the research effort to identify critical biological targets for determining radiation quality has been the underscoring of the importance of target sizes of a few nanometres in diameter. This outcome in turn has generated a considerable effort to develop novel experimental methods that can measure energy deposition on this dimensional scale. One such method, known as the variance technique, has been particularly valuable in this regard and will receive particular attention.

1.2 EXPERIMENTAL MICRODOSIMETRY AND RADIATION METROLOGY

At the same time as the growing realisation that relevant biological target sizes were likely to be much smaller than the site sizes typically used in classical microdosimetry, work continued for dosimetry purposes in examining the properties of tissue-equivalent proportional counters (TEPCs) that simulated site-sizes of a few micrometers. This research gradually established the TEPC as an essential instrument for use in radiation metrology where advantage could be taken of both the dosimetric properties and spectroscopic properties of these devices, which enabled, for example, the estimation of quality factors and the analysis and unscrambling of mixed radiation fields. The use of the TEPC in radiation metrology has assumed sufficient importance over the years to become a field of research in its own right and independent of the ongoing quest for the most appropriate target size and means of quantifying radiation quality. This aspect of experimental microdosimetry will be examined and discussed in considerable detail.

1.3 TEXTS AND DOCUMENTATION SOURCES ON MICRODOSIMETRY

Microdosimetry is an evolving field of research in radiation science and although a fundamental topic, considered by many to underlie all of radiation science, it is nevertheless still a niche field of study and practice. Consequently, many standard textbooks on radiation science and dosimetry describe microdosimetry only in a very cursory manner or not at all. Certainly the concepts and methods of microdosimetry have been documented in the past as chapters in specialist books and reports (such as Kellerer, 1985) as well as in the proceedings of the numerous symposia on microdosimetry that began in 1967. These texts are becoming accessible with an increasing degree of difficulty as time passes. As mentioned earlier, the work of Rossi and Zaider is perhaps the most comprehensive standalone text on microdosimetry and still an essential read for a complete understanding of the field; however, at the time of this writing it is nearly 20 years since its publication. With this text we have striven above all to provide the interested reader with an up-to-date practical guide to the methods of experimental microdosimetry in the hope that it will assist in the continued development and application of the field by a new generation of researchers and radiation measurement scientists and engineers.

1.4 ORGANISATION OF MATERIAL AND TEXT

The organisation of the material in this volume is in two parts. We begin with a chapter dealing with quantities and units used in microdosimetry followed by a chapter describing the various experimental techniques employed. Following this review of fundamentals, the second part, also consisting of two chapters, deals with the principal applications of experimental microdosimetry in radiation protection and radiation therapy. Newcomers to the field will benefit from reading the text sequentially. For those who already have some familiarity with microdosimetry we hope we have provided a useful reference text that can be used as a research resource and as a starting point for discussions with students and colleagues.

CHAPTER 2

Quantities in Experimental Microdosimetry

2.1 INTRODUCTION

Radiation interacts with matter in a stochastic manner that can be illustrated experimentally, for instance, using a cloud chamber. In practical dosimetry, this stochastic behaviour is overcome by working with dosimetry quantities that represent averages and by using detectors, such as ionisation chambers, which are large enough to ensure that fluctuations in energy deposition are small. These average quantities are well established in classical dosimetry and are utilised in numerous dosimetry protocols.

The field of microdosimetry began with concepts developed by Prof. H. H. Rossi in the 1950s at Columbia University in New York (Rossi, 1959; Rossi and Rosenzweig, 1955a,b; Rossi et al., 1961). The fundamental idea underlying microdosimetry is to explore whether taking the stochastic nature of radiation interaction into account will improve our understanding of the relationship between physics and results observed in radiation biology and connected fields such as radiation therapy and radiation protection. The statistical distributions of the energy deposited were originally available only from measurements but today can be obtained computationally using Monte Carlo track structure calculations as well as some condensed Monte Carlo codes (Nikjoo and Liamsuwan, 2014; Nikjoo et al., 2012). Both tools are essential for a correct understanding of the physics; however, in this textbook the focus is on experimental methods, in particular those which have been found to be useful in radiation therapy and radiation protection applications.

The main stochastic quantities used in microdosimetry are presented in this chapter and their relationship to their corresponding average quantities is emphasised. These quantities are not meant to replace existing average quantities but to complement them with information about the probability for the deposited energy to reach a certain value within a specified volume when an ionising particle interacts with the volume. The definitions used are those presented by the International Commission on Radiation Units & Measurements (ICRU) in its Reports 36 (ICRU, 1983) and 85a (ICRU, 2011). For a complete set of definitions and notes, the reader is referred to these two ICRU documents. The notation has changed somewhat between the two reports and the notation from ICRU 85a has been adopted in this work. Before discussing these quantities, a few words concerning statistical functions are in order.

2.1.1 Probability Distributions

In definitions of stochastic microdosimetric quantities probability distributions, their derivatives, and mean values, are indispensable. The probability distribution function, $F(x)$, defines the probability that the random variable $\underline{x}$ is equal to or less than the value x. $F(x)$ can take any value between 0 and 1. The probability density, $f(x)$, is the derivative of $F(x)$ and is the probability of observing an event within the interval x and $x + dx$. It follows that

$$\int_0^\infty f(x)dx = 1 \tag{2.1}$$

The expectation value is

$$\bar{x} = \int_0^\infty xf(x)dx \tag{2.2}$$

To give an example of the use of these probability distributions, suppose we wish to test a threshold hypothesis that states events with a size below or equal to 7 ($x \leq 7$) will not be able to produce a particular effect, but larger events will. What is the probability that we will observe the effect for one random event? With $F(x)$ defined in the left panel of Figure 2.1, the answer is given by $1 - F(7) = 0.35$. The probability of an event having a size of exactly 7 is given by the probability density function, $f(x)$ and again, from the left panel of Figure 2.1a, $f(7) = 0.15$. The probability density function, $f(x)$, is usually called the frequency distribution because it relates to the number of occurrences of events of each size.

If instead of a threshold the probability of the effect is increasing linearly with the size of the event, a size-weighted distribution becomes of interest. In this case, the number of events in each size interval is multiplied by the size of the event, and, after normalization, we arrive at a probability distribution function, $D(x)$. This distribution defines the probability that the *weight* (frequency times size) of the random variable $\underline{x}$ is equal to or less than the value x. $D(x)$ can take any value between 0 and 1. The probability density, $d(x)$, is the derivative of $D(x)$, and

$$\int_0^\infty d(x)dx = 1 \tag{2.3}$$

In microdosimetry, size-weighted distributions such as $d(x)$ are called dose distributions. The expectation value of $d(x)$ is

$$\bar{x}_D = \int xd(x)dx = \frac{\int x^2 f(x)dx}{\bar{x}} \tag{2.4}$$

The four functions may be used with any of the stochastic quantities defined below.

Continuing with our example, we may like to know the fraction of the total *weight* by events larger than 7. Figure 2.1b presents the two functions $F(x)$ and $D(x)$. The *weight* by events larger than 7 is $[1 - D(7)] = 0.62$. Thus, 34% of the events $[1 - F(7)]$ cause 62% of the weight.

It follows that the mean value, $\bar{x}_D$, represents this distribution exactly on condition that the probability of the effect is increasing linearly with the size of the event. If the relation is more complex, $\bar{x}_D$ will become more or less good as an approximation for the whole distribution.

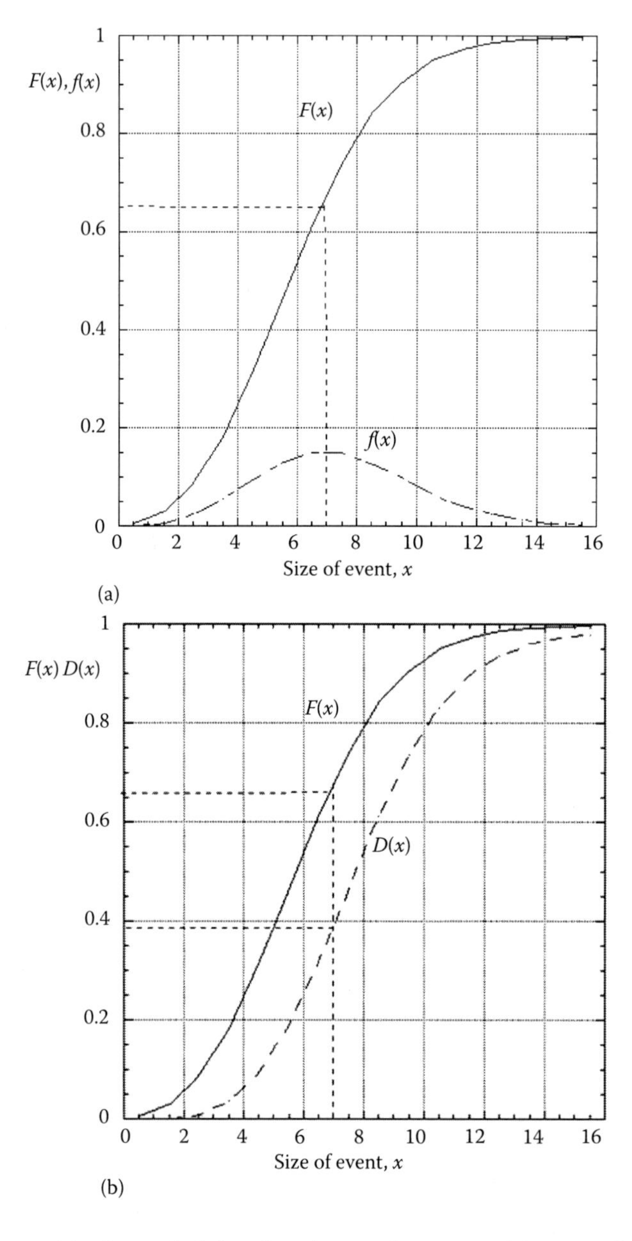

FIGURE 2.1 (a) The probability distribution function, $F(x)$, varies from 0 to 1. For example, the probability to find $x \leq 7$ is $F(7) = 0.65$. The frequency distribution, $f(x)$, which is the derivate of $F(x)$, is the probability to find x in the limited interval x to $x + dx$, in the example $f(7) = 0.15$. The $f(x)$ distribution is by definition normalised to one. (b) $F(x)$ and $D(x)$. The dotted lines mark the probability of having events of size ≤ 7, $F(7) = 0.65$, and the size-weighted fraction delivered by those events, $D(7) = 0.38$. See text.

2.1.2 The Track Length Distribution

Before coming to the definitions of the quantities for the energy deposited by ionising particles, it is useful to define a quantity characterising the track laid down by these particles. A charged particle will pass through a volume along a random chord length, ℓ. The energy deposited in the volume will therefore depend on the shape of the volume. As the choice of route is random, it will be described by a probability distribution function. When a body is exposed to a uniform isotropic fluence of infinite straight lines, this randomness is called μ-randomness (Kellerer, 1984). As an example, the distribution function for a sphere with diameter d is

$$F(\ell) = 1 - \left(\frac{\ell}{d}\right)^2 \tag{2.5}$$

As before, the distribution function $F(\ell)$ can take any value between 0 and 1. $F(\ell)$ is the probability that the random variable $\underline{\ell}$ is equal to or less than the value ℓ. The probability density distribution for the sphere is

$$f(\ell) = \frac{2\ell}{d^2} \quad 0 \leq \ell \leq d \quad f(\ell) = 0 \text{ when } \ell > d \tag{2.6}$$

and

$$F(\ell) = \int_0^d f(\ell)d\ell = 1 \tag{2.7}$$

The mean chord length is

$$\bar{\ell} = \int_0^d \ell f(\ell)d\ell \tag{2.8}$$

The mean chord length enters in the definition of the lineal energy, a principal quantity used in experimental microdosimetry. For a convex body randomly intersected by chords, the mean length of the chords is given by $\bar{\ell} = \frac{4V}{S}$, where V is the volume and S is the surface area (ICRU, 1983). Thus for a sphere the mean chord length is 2/3 of the sphere diameter; this is also true for a cylinder where the height of the cylinder, h, is equal to its diameter, d. For cylinders in general the formula for $\bar{\ell}$ is

$$\bar{\ell} = \frac{2rh}{r + h} \tag{2.9}$$

where r is the radius and h is the height of the cylinder. In Figure 2.2 (Hogeweg, 1978) the chord length distributions for three different volumes are derived from Monte Carlo calculations. All three volumes are designed to have the same mean chord lengths, 0.48 cm. The distribution function is triangular for the sphere, as predicted by Equation 2.6.

Knowing the relationship between $\bar{\ell}$ and the dimensions of the geometrical volumes, we can calculate the height and diameter of the three example volumes given in Figure 2.2.

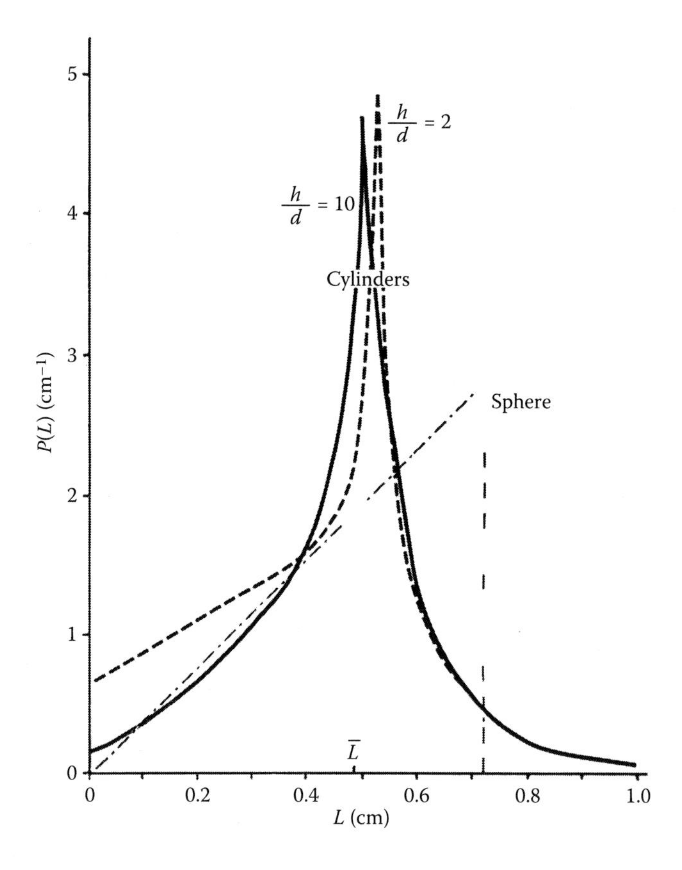

FIGURE 2.2 Chord length distributions, for cylinders with elongation factors, h/d, equal to 2 and 10, and for a sphere, h is the cylinder height and d is its diameter. The mean track length is identical for the three bodies and equal to 0.48 cm. (From Hogeweg B, Microdosimetric measurements and some applications in radiobiology and radiation protection, PhD thesis, Radiobiological Institute of the Organization for Health Research TNO, Rijswijk, the Netherlands, 1978.) Observe that in the figure the notation for the chord length is L instead of ℓ. See text for comments.

a. The sphere diameter is calculated directly from $\bar{\ell} = \frac{2d}{3}$ and with $\bar{\ell} =$ 0.48 cm, $d = 0.72$ cm.

b. For the cylinders h/d is 2 and 10 (see figure caption). Inserting $h = 2d$ and $h = 10d$ respectively and $\bar{\ell} = 0.48$ cm into Equation 2.9, we obtain the results $d = 0.6$ cm and $h = 1.2$ cm and $d = 0.5$ cm and $h = 5$ cm. From this analysis, we see that for very elongated cylinders $\bar{\ell}$ is close in value to the cylinder diameter (0.48 cm and 0.5 cm respectively).

2.1.3 Stochastic and Average Quantities

Stochastic and (non-stochastic) average quantities will be introduced together in an effort to make the relation between the two clear. We start with *dose quantities* and continue with *linear quantities*. The latter are used to describe charged particle tracks. In microdosimetry the most basic of all the quantities is the stochastic quantity '*energy deposit* in a single interaction'. This is followed by the stochastic quantities *energy imparted* and *specific energy* relevant to the amount of energy deposited in a defined limited volume. The average of specific energy is related to the non-stochastic quantity absorbed dose and this relationship will be discussed and demonstrated with a few examples. Linear energy transfer, lineal energy and in particular the relation between the two will be treated separately following the discussion of the dose quantities.

The term *single* energy deposition event or sometimes just single event is often used in microdosimetry texts and stands for the energy imparted by correlated charged particles, when passing through a specified volume. Correlated particles are, for instance, a charged particle and its delta electrons or the primary and secondary particles in nuclear reactions. If, for example, delta electrons from an uncorrelated charged particle deposit energy as well, the event is then called a multi-event. Figure 2.3a and b give examples of these two situations. It is important that energy *is* deposited, otherwise, the event would not of course be experimentally observable. The relevance of this definition of an event comes, most likely, from the early assumption by radiation biologists that if no energy were deposited there would be no biological effect. Today it is known that signalling may occur between cells and the situation is rather more complex than was originally thought.

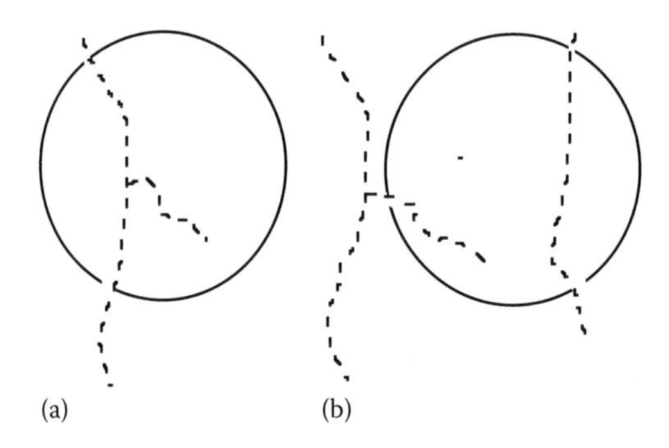

FIGURE 2.3 (a, b) A particle interacts with atoms at *transfer points* (marked with a dash) when passing through matter. In (a) all transfer points are correlated, that is, they originate within the same particle track, and this is termed a single event. In (b) the transfer points are due to two independent particles, and this circumstance is termed a multi-event. Whether the energy imparted inside a volume is due to one or multiple events can be predicted only by statistical means.

The use of *event* rather than particle track emphasises that in microdosimetry it is not necessarily the energy imparted in the main charged particle track that is relevant. In very small volumes, a delta electron or any other secondary charged particle may be solely responsible for the imparted energy and may cause the effect being studied.

In the definitions of the stochastic quantities only the probability density functions, $f(x)$ and $d(x)$, are mentioned. They are related to the probability distributions, $F(x)$ and $D(x)$, (Section 2.1.1), which vary from 0 to 1. The integrals of the probability density functions, $f(x)$ and $d(x)$, are equal to 1. In the following, these two distributions will be referred to as the frequency and dose distribution functions, respectively.

2.2 DOSE-RELATED QUANTITIES

The most fundamental quantity in microdosimetry is the *energy deposited* in a single interaction. The *energy imparted* is the sum of all energy depositions in a defined volume, and finally the *specific energy* is the energy imparted divided by the mass of that volume. The size and form of the volume is a user definable variable.

2.2.1 Energy Deposit, ε_i

Energy deposit, ε_i, is the energy deposited in a single interaction, i, and

$$\varepsilon_i = \varepsilon_{\text{in}} - \varepsilon_{\text{out}} + Q \tag{2.10}$$

where ε_{in} is the energy of the incident ionising particle (excluding rest energy), ε_{out} is the sum of the energies of all charged and uncharged ionising particles leaving the interaction (excluding rest energy) and Q is the change in the rest energies of the nucleus and of all elementary particles involved in the interaction ($Q > 0$: decrease of rest energy; $Q < 0$: increase of rest energy).

Unit: J (although most often expressed in terms of electron-volts, eV).

2.2.2 Energy Imparted, ε

The energy imparted, ε, to the matter in a given volume is the sum of all energy deposits in the volume; thus

$$\varepsilon = \sum_i \varepsilon_i \tag{2.11}$$

where the summation is performed over all energy deposits, ε_i, as illustrated in Figure 2.4. Unit: (J). The quantity may also be expressed in eV. ε is a stochastic quantity and can be used both for single- and multi-event situations.

2.2.3 Specific Energy, *z*

The specific energy (imparted), z, is the quotient of ε divided by m, where ε is the energy imparted by ionising radiation to matter in a volume of mass m; thus

$$z = \frac{\varepsilon}{m} \tag{2.12}$$

Unit: J kg^{-1}, where the special name for the unit of specific energy is gray (Gy).

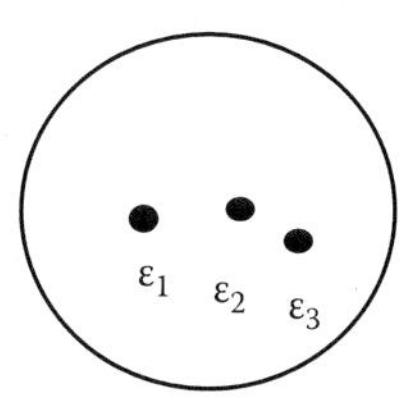

FIGURE 2.4 Schematic presentation of energy deposits ε_1, ε_2 and ε_3 at three transfer points within the specified volume of mass m and mean chord length $\bar{\ell}$. Energy imparted is $\varepsilon = \varepsilon_1 + \varepsilon_2 + \varepsilon_3$; specific energy $z = (\varepsilon_1 + \varepsilon_2 + \varepsilon_3)/m$ and lineal energy $y = (\varepsilon_1 + \varepsilon_2 + \varepsilon_3)/\bar{\ell}$.

As ε is a stochastic quantity, it follows that specific energy is also a stochastic quantity. Specific energy is defined for both a single event and multi-events.

The frequency function, $f(z)$, includes a discrete component (a Dirac delta function) at $z = 0$ for the probability of no energy deposition. The expectation value

$$\bar{z} = \int_{0}^{\infty} z f(z) dz \tag{2.13}$$

is called mean specific energy and this average is a non-stochastic quantity. As we will see later, this mean value will be equal to the absorbed dose. When the specific energy is deposited by single events the notation for the probability density is $f_s(z)$. This distribution is called the *single-event* frequency distribution of z and its expectation value is called the single-event frequency-mean specific energy. It is defined as

$$\bar{z}_s = \int_{z_{\min}}^{z_{\max}} z f_s(z) dz \tag{2.14}$$

where $\bar{z}_s$ is a non-stochastic quantity.

The ratio of the mean value of the specific energy of the multi-event distribution, $\bar{z}$, to that of the single-event distribution, $\bar{z}_s$, defines the mean number of energy deposition events, $\bar{n}$, that have contributed to $\bar{z}$ and

$$\bar{n} = \frac{\bar{z}}{\bar{z}_s} \tag{2.15}$$

The number of energy deposition events is distributed randomly and may be described by a Poisson distribution, as discussed in the text that follows.

An important distribution for single-energy deposition events is the dose distribution function, $d_s(z)$, and the expectation value is given by

$$\bar{z}_{D,s} = \int_{z_{\min}}^{z_{\max}} z d_s(z) dz \tag{2.16}$$

and is called the single-event dose-mean specific energy. $\bar{z}_{D,s}$ is a non-stochastic quantity.

The single-event dose-mean specific energy can also be written as

$$\bar{z}_{D,s} = \frac{1}{\bar{z}_s} \int_{z_{\min}}^{z_{\max}} z^2 f_s(z)\,dz \tag{2.17}$$

or

$$\bar{z}_{D,s} = \frac{\overline{z_s^2}}{\bar{z}_s} \tag{2.18}$$

2.2.4 Absorbed Dose, *D*

The absorbed dose, *D*, is the quotient of $d\bar{\varepsilon}$ by dm, where $d\bar{\varepsilon}$ is the mean energy imparted by ionising radiation to matter of mass dm; thus

$$D = \frac{d\bar{\varepsilon}}{dm} \tag{2.19}$$

Unit: J kg^{-1}. The special name for the unit is gray (Gy).

As $d\bar{\varepsilon}$ is the mean energy imparted, the absorbed dose becomes an average or non-stochastic quantity.

2.2.5 Relations between $\bar{z}$ and *D*

Before continuing with other quantities it is useful to derive the relation between absorbed dose, *D*, and the mean specific energy, $\bar{z}$. With

$$z = \frac{\varepsilon}{m} \tag{2.20}$$

it follows that

$$\bar{z} = \int_0^\infty z f(z)\,dz = \int_0^\infty \frac{\varepsilon f(\varepsilon)\,d\varepsilon}{m} = \frac{\bar{\varepsilon}}{m} \tag{2.21}$$

Usually dm in the definition of absorbed dose, *D*, is supposed to be an infinitesimally small volume and *D* is considered to be defined at a point. However, in practice the absorbed dose must be measured with a detector of a definite dimension filled with a certain amount of matter; otherwise no interaction will take place and no energy will be deposited. Thus the average absorbed dose in that volume becomes

$$\bar{D} = \frac{\bar{\varepsilon}}{m} = \bar{z} \tag{2.22}$$

Usually the difference between D and $\bar{D}$ is ignored. The mean event number, $\bar{n}$, is thus

$$\bar{n} = \frac{\bar{z}}{\bar{z}_s} = \frac{D}{\bar{z}_s} \tag{2.23}$$

The mean number of events per unit absorbed dose is called the event frequency, $\Phi^*(0)$, and

$$\Phi^*(0) = \frac{\bar{n}}{D} = \frac{1}{\bar{z}_s} \tag{2.24}$$

The variance for any distribution is the second moment minus the square of its first moment. For the *single*-event distribution this variance is

$$V_s = \overline{z_s^2} - \bar{z}_s^2 \tag{2.25}$$

The *relative* variance of the distribution is then

$$V_{s,\,\mathrm{rel}} = \frac{V_s}{(\bar{z}_s)^2} = \frac{\bar{z}_{Ds}}{\bar{z}_s} - 1 \tag{2.26}$$

For the *multi*-event distribution the variance V_n becomes

$$V_n = \overline{z^2} - \bar{z}^2 \tag{2.27}$$

and its relative variance is

$$V_{n,\,\mathrm{rel}} = \frac{V_n}{(\bar{z})^2} = \frac{\overline{z^2}}{\bar{z}^2} - 1 \tag{2.28}$$

This variance may also be derived from $V_{s,\mathrm{rel}}$ (Equation 2.26) (Kellerer, 1985). The variance of the sum of independent random variables is equal to the sum of their variances. If the specific energy was due to exactly n events and there is no fluctuation connected with n and $z = n\bar{z}_s$ then the variance of the multi-event distribution is $V_n^{'} = nV_s$. The relative variances are thus related as

$$V_{n,\mathrm{rel}}^{'} = \frac{V_n^{'}}{\bar{z}^2} = \frac{V_{s,\,\mathrm{rel}}}{n} = \frac{1}{n}\left(\frac{\bar{z}_{Ds}}{\bar{z}_s} - 1\right) \tag{2.29}$$

However, the number of events varies as well, and as this number for constant dose rate has a Poisson distribution its variance is $1/\bar{n}$ and has to be added. Thus

$$V_{n,\mathrm{rel}} = \frac{1}{\bar{n}}\left[\frac{\bar{z}_{D,s}}{\bar{z}_s} - 1\right] + \frac{1}{\bar{n}} = \frac{\bar{z}_{D,s}}{\bar{z}} = \frac{\bar{z}_{D,s}}{D} \tag{2.30}$$

This relation is used in the variance method, an experimental method for measuring $\bar{z}_{D,s}$ and is dealt with in detail in Chapter 2. The second moment now becomes (Equations 2.28, 2.29 and 2.31)

$$\overline{z^2} = V_n + \bar{z}^2 = \bar{z}_{D,s}\bar{z} + \bar{z}^2 \tag{2.31}$$

or

$$\overline{z^2} = \bar{z}_{D,s}D + D^2 \tag{2.32}$$

The relationship expressed in Equation 2.32 will be used again in Chapter 3, when the theory of dual radiation action is discussed.

2.2.6 Poisson Distribution of the Number of Events and the Variance in $\bar{z}$

The Poisson distribution determines the probability that a mean specific energy or absorbed dose, D, is due to exactly ν events and is given by

$$p(\nu) = \frac{\bar{n}^{\nu} e^{-\bar{n}}}{\nu!} \tag{2.33}$$

The probability that D is due to ν events is calculated for three absorbed dose values of 10 μGy, 1 mGy and 6 mGy after irradiation with ^{60}Co γ. The mean specific energy of the single-event distribution is $\bar{z}_s$ = 0.87 mGy, when the volume diameter is 8 μm. The results are also shown in Figure 2.5 and Table 2.1.

The dose values listed at the bottom of the table are obtained as $\bar{z} = \bar{z}_s\bar{n} \approx D$ or as

$$\bar{z} = \sum_{\nu=0}^{\nu_{\max}} \nu p(\nu)\bar{z}_s \tag{2.34}$$

If ν and $p(\nu)$ from Table 2.1 are inserted in Equation 2.34 for the absorbed dose 10 μGy and 1 mGy respectively, $\bar{z}$ becomes $\bar{z} = \sum_{\nu=0}^{\nu_{\max}} \nu p\ (\nu)\bar{z}_s =$ $(0 \times 0.99 + 1 \times 0.012)0.87 \approx 0.010$ mGy

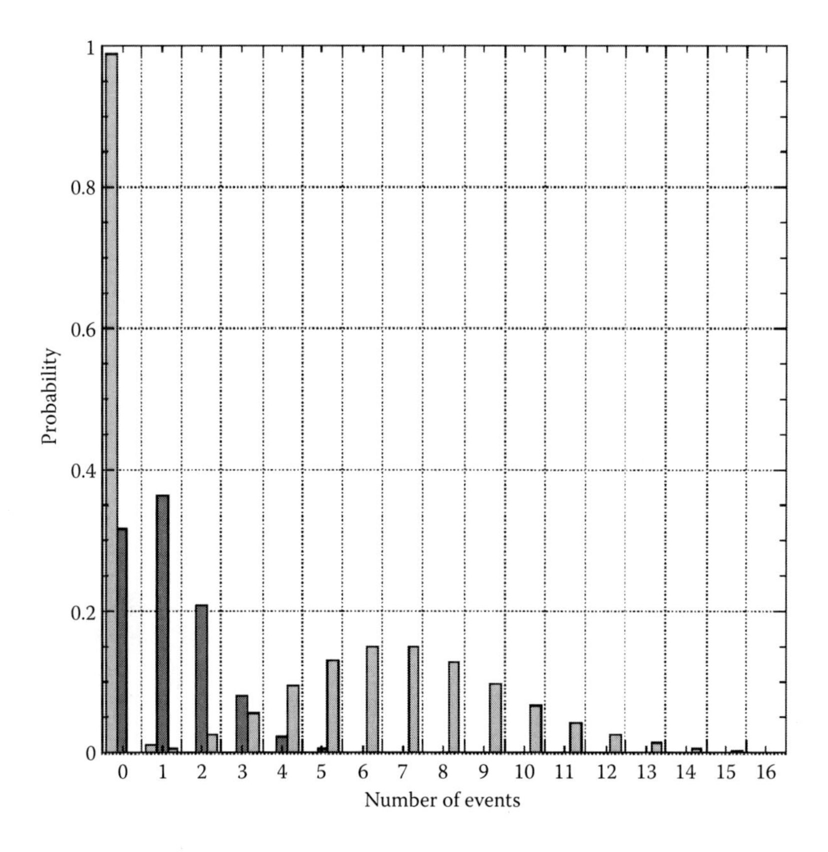

FIGURE 2.5 **(See colour insert.)** The probability that D is caused by ν events when D = 10 μGy (blue columns), D = 1 mGy (red columns) and D = 6 mGy (green columns) in a volume of 8 μm diameter (approximately the diameter of mammalian cell nucleus). The radiation beam is ^{60}Co γ.

$$\bar{z} = \sum_{\nu=0}^{\nu_{\max}} \nu p(\nu)\bar{z}_s = (0 \times 0.317 + 1 \times 0.364 + 2 \times 0.209 + 3 \times 0.08 + 4 \times 0.023 + 5 \times 0.005)0.87 \approx 1.0 \text{ mGy}$$

The very large difference between $\bar{z}_s$ (0.87 mGy) and D at 10 μGy is due to the large number of charged particles not depositing energy in the volume (sometimes called zero events). Zero events are not included in the number of events, when $\bar{z}_s$ is calculated, contrary to the situation for D. Another conclusion that can be drawn from this calculation is that it is

TABLE 2.1 Probability for the Absorbed Dose to Be due to a Specific Number of Events

	v	$p(v)$ D = 10 μGy	$p(v)$ D = 1 mGy	$p(v)$ D = 6 mGy
$\bar{n} = D/\bar{z}_s$		**0.012**	**1.15**	**6.90**
	0	0.99	0.317	0.00
	1	0.01	0.364	0.007
	2	0.00	0.209	0.024
	3	0.00	0.080	0.055
	4	0.00	0.023	0.095
	5		0.005	0.131
	6			0.151
	7			0.149
	8			0.128
	9			0.0985
	10			0.0679
	11			0.0426
	12			0.0245
	13			0.0130
$\bar{z} = \sum_{\nu=0}^{\nu_{max}} \nu p(\nu)\bar{z}_s$		10 μGy	1.00 mGy	6.00 mGy

Note: The probability that D is due to ν events is calculated for three absorbed dose values of 10 μGy, 1 mGy, and 6 mGy after irradiation with ^{60}Co γ radiation. The mean specific energy of the single-event distribution is 0.87 mGy, when the volume diameter is 8 μm. The results are also shown in Figure 2.5.

never possible to state the exact number of events that deposit the energy – only the probability for a certain number. As will be underlined later, single-event distributions are measured under the condition that multi-events can be ignored. If we are interested in the mean specific energy in just those volumes where at least one event has occurred, $\bar{z}_h$, it is calculated as

$$\bar{z}_h = \frac{\bar{z}}{1-p(0)} = \frac{D}{1-p(0)} \tag{2.35}$$

At $D = 1$ mGy $\bar{z}_h = \frac{1}{1-0.317} = 1.46$ mGy

The example presented earlier for a volume of 8 μm diameter corresponds to the situation for a typical mammalian cell nucleus. The dose values chosen correspond to absorbed dose values relevant in the field of radiation protection. The dose 10 μGy is not uncommon as the dose received during one hour of radiological work, while 1 mGy may correspond to the annual natural dose from the environment (excluding radon and medical examinations). A dose of 6 mGy or more during one year is usually a condition for a radiological worker to wear a personnel dosemeter. Table 2.1 demonstrates that at 10 μGy 99% of the cell nuclei are likely not to be hit at all. At 1 mGy it is likely that almost 70% of the nuclei are hit by one or more events. At 6 mGy all nuclei are hit and the probability to be hit by five events or more is approximately 80%. The probability the nucleus will be hit by 10 or more events is 15%. As a consequence, the specific energy will vary strongly between the individual nuclei hit.

The dose mean specific energy, $\bar{z}_{D,s}$, of a *single*-event distribution is related to the relative variance, $V_{n,\text{rel}}$, of the multi-event distribution of the specific energy (Equation 2.30); thus there is always a variance component in the dose that is related to the single-event distribution, independent of the magnitude of the value of the absorbed dose, D. The following example will illustrate this: We would like to know the relative standard deviation, σ_{rel} at $D = 6$ mGy caused by microdosimetric sources in a tissue equivalent volume with 8 μm diameter when irradiated with a ^{60}Co γ source. From experiments we know that $\bar{z}_{D,s} = 4.1$ mGy. From Table 2.1 we find that the probability for a zero event to occur is small and we can apply Equation 2.31. The relative standard deviation of this multi-event distribution is

$$\sigma_{\text{rel}} = \pm\sqrt{V_{\text{rel}}} = \sqrt{\frac{\bar{z}_{D,s}}{D}} \qquad \text{or} \qquad \sigma_r = \pm\sqrt{\frac{4.1}{6}} = \pm 0.83$$

The mean value of the specific energy and its standard deviation are thus 6 ± 5 mGy. If the multi-event distribution includes zero-events (some volumes are without energy deposits) the relative standard deviation of the specific energy in those volumes in which *at least one event* has occurred is (ICRU, 1983)

$$\frac{\sigma_{z,h}}{\bar{z}_h} = \left\{\left(\frac{\bar{z}_{D,s}}{D}\right) - \left(\frac{\bar{z}_{D,s}}{D} + 1\right)e^{-D/\bar{z}_s}\right\}^{1/2} \tag{2.36}$$

So far, only two averages of specific energy distributions have been used, but they have been useful and sufficient to point to important features of the way the absorbed dose is distributed amongst small volumes. This information has been easy to extract from the specific energy averages; much more detailed information becomes available when the whole specific energy single-event distribution is available.

2.3 TRACK-RELATED QUANTITIES

2.3.1 Lineal Energy, y

The lineal energy, y, is the quotient of ε_s divided by $\bar{\ell}$, where ε_s is the energy imparted to matter in a given volume by a *single* energy-deposition event and $\bar{\ell}$ is the mean chord length in that volume; thus

$$y = \frac{\varepsilon_s}{\bar{\ell}} \tag{2.37}$$

Unit: joule per metre (J m^{-1}). The numerator, ε_s, may be expressed in eV; hence y may be expressed in multiples and submultiples of eV and m, for example, in keV μm^{-1}. The quantity y is a stochastic quantity; its frequency function is $f(y)$ and its expectation value

$$\bar{y} = \int_0^\infty y f(y) dy \tag{2.38}$$

is called frequency-mean lineal energy and this average is a non-stochastic quantity. The expectation value of the dose distribution, $d(y)$,

$$\bar{y}_D = \int_0^\infty y d(y) dy \tag{2.39}$$

is called the dose-mean lineal energy. $\bar{y}_D$ is a non-stochastic quantity. The relation between $d(y)$ and $f(y)$ is

$$d(y) = \frac{y}{\bar{y}} f(y) \tag{2.40}$$

and

$$\bar{y}_D = \frac{1}{\bar{y}} \int_0^\infty y^2 f(y) dy \tag{2.41}$$

The distributions $f(y)$ and $d(y)$ are independent of absorbed dose but are dependent on the size and shape of the volume in which the energy is imparted.

2.3.2 Linear Energy Transfer, L

The linear energy transfer (LET) or restricted linear electronic stopping power, L_Δ, of a material for charged particles of a given type and energy, is the quotient of dE_Δ divided by dl, where dE_Δ is the mean energy lost by a charged particle due to electronic interactions in traversing a distance dl, minus the sum of the kinetic energies of all the electrons released with kinetic energies in excess of Δ; thus

$$L_\Delta = \frac{dE_\Delta}{dl} \tag{2.42}$$

Unit: J m^{-1}. Other units such as keV μm^{-1} can be used.

Because L_Δ expresses the energy balance between the energy lost by the primary charged particles in interactions with electrons, along a distance dl, minus the energy carried away by secondary electrons having initial kinetic energies greater than Δ, it is a quantification of the energy that is considered to be 'locally transferred'. 'Locally' in this context means a small volume surrounding the primary charged particle. When no cut-off is applied the LET is denoted as L_∞.

As a charged particle loses energy along its track, its LET will change and if $t(L_\Delta)dl$ is the fraction of the track with LET between L_Δ and $L_\Delta + \Delta L$ the track average LET is defined as

$$\bar{L}_\Delta = \int L_\Delta t(L_\Delta)dl \tag{2.43}$$

If $t(L_\Delta)$ is weighted with the energy lost in each track length interval and the distribution is normalised, $d(L_\Delta)dl$ will be the fraction of the dose that is deposited between L_Δ and $L_\Delta + dL$. The average quantity defined by

$$\bar{L}_{\Delta,D} = \int L_\Delta d(L_\Delta)dl \tag{2.44}$$

is called the dose average LET (ICRU, 1970).

2.3.3 Relationship between L and y

LET and lineal energy are both quantities that characterise particle tracks. While LET is useful for defining the radiation quality of a radiation field itself, lineal energy is useful when the radiation quality needs to be characterised in small volumes of irradiated matter. ICRU has given the following conditions to be fulfilled for lineal energy and linear energy transfer to be equal: If particles of a fixed LET, L_0, are crossing a sphere with diameter, *d*, and the ranges of the particles are long compared to the diameter of the sphere and if energy is lost continuously along the particle track and this energy is completely deposited inside the sphere, then $\bar{y} = L_0$ and $\bar{y}_D = 9L_0/8$ (ICRU, 1983). As will be seen, in large volumes we can expect LET and lineal energy to be numerically close; however, in nanometre volumes this numerical equality breaks down.

In Figure 2.6a, b and c, a small part of a simulated track of a 300 MeV u^{-1} carbon ion (Hultqvist, 2011) is intercepted by a spherical volume of 1 μm diameter. The particle may take any possible chord length through the

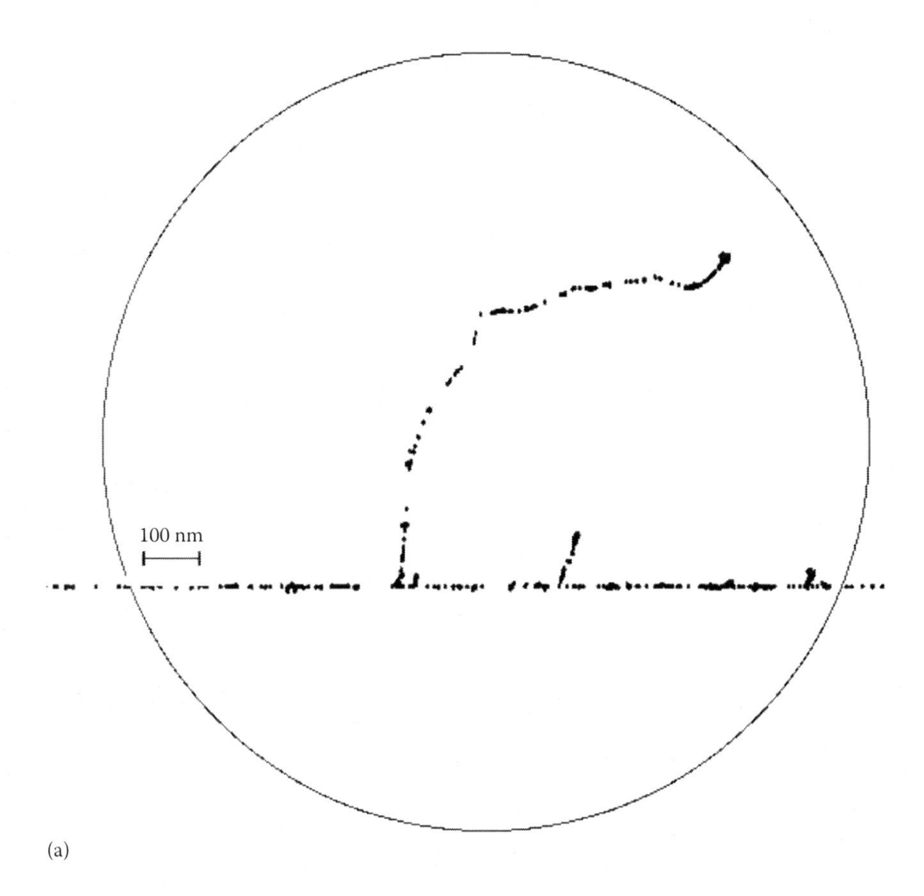

FIGURE 2.6 (a) Schematic description of the lineal energy. A carbon ion of 300 MeV u^{-1} is passing through a sphere of about 1000 nm diameter. The Figures 2.6a, b and c illustrate three possible chords the particle may follow. In (a) delta particles will not contribute to the lineal energy separately (particle track simulated by Hultqvist, 2011). *(Continued)*

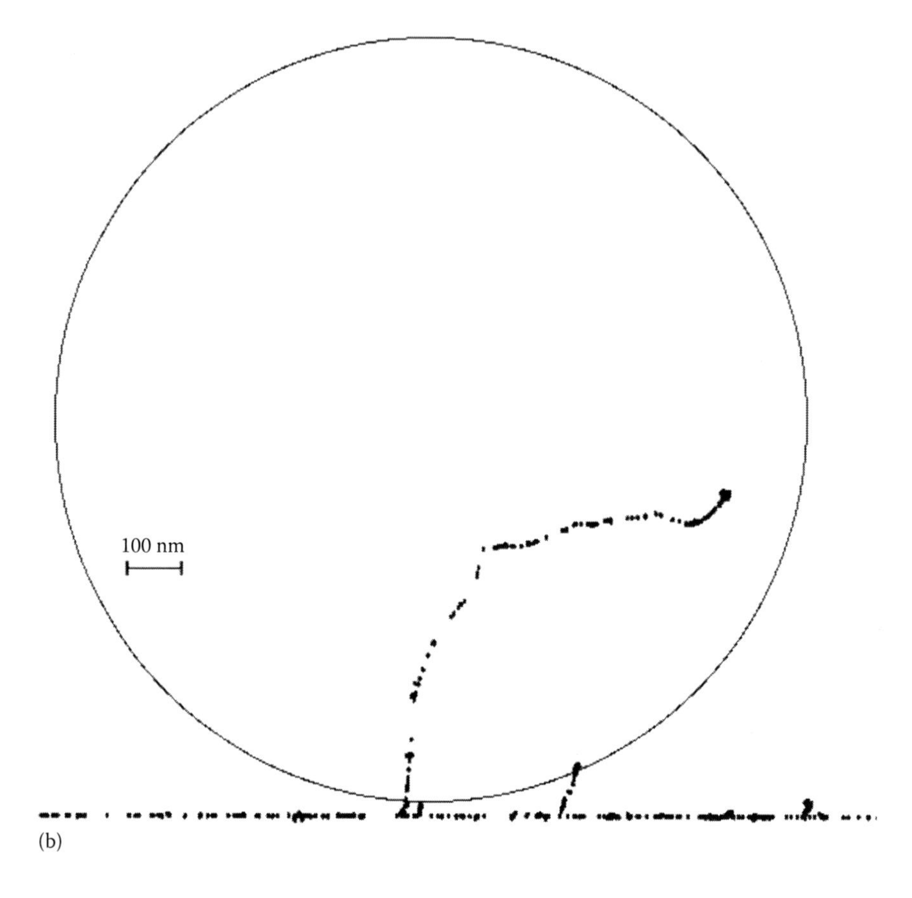

FIGURE 2.6 (CONTINUED) (b) The same particle track as in (a) but the main track passes outside the sphere and a delta particle is entering it. (*Continued*)

volume and the figures show three possibilities. In Figure 2.6a both the main track and some delta electrons will deposit energy; in Figure 2.6b only delta electrons will do so and in Figure 2.6c only the charged particle track core deposits the energy. In Figure 2.6d, L_∞ is depicted. The ratio of energy transferred, dE, by the charged particle when passing through the layer dl constitutes L_∞. While the way the energy imparted in Figures 2.6a and d looks quite similar, the manner in which the energy is imparted in Figures 2.6b and c is clearly different from that shown in Figure 2.6d. The difference becomes more evident in Figure 2.6e. Here the energy deposited by the same carbon ion track but intercepted by volumes of 100 nm diameter is pictured. Instead of exemplifying 24 possible paths through a volume with 24 figures, Figure 2.6e includes 24 possibilities for the track to deposit energy in the volume. The similarity between y and L_∞ is now lost. Also in

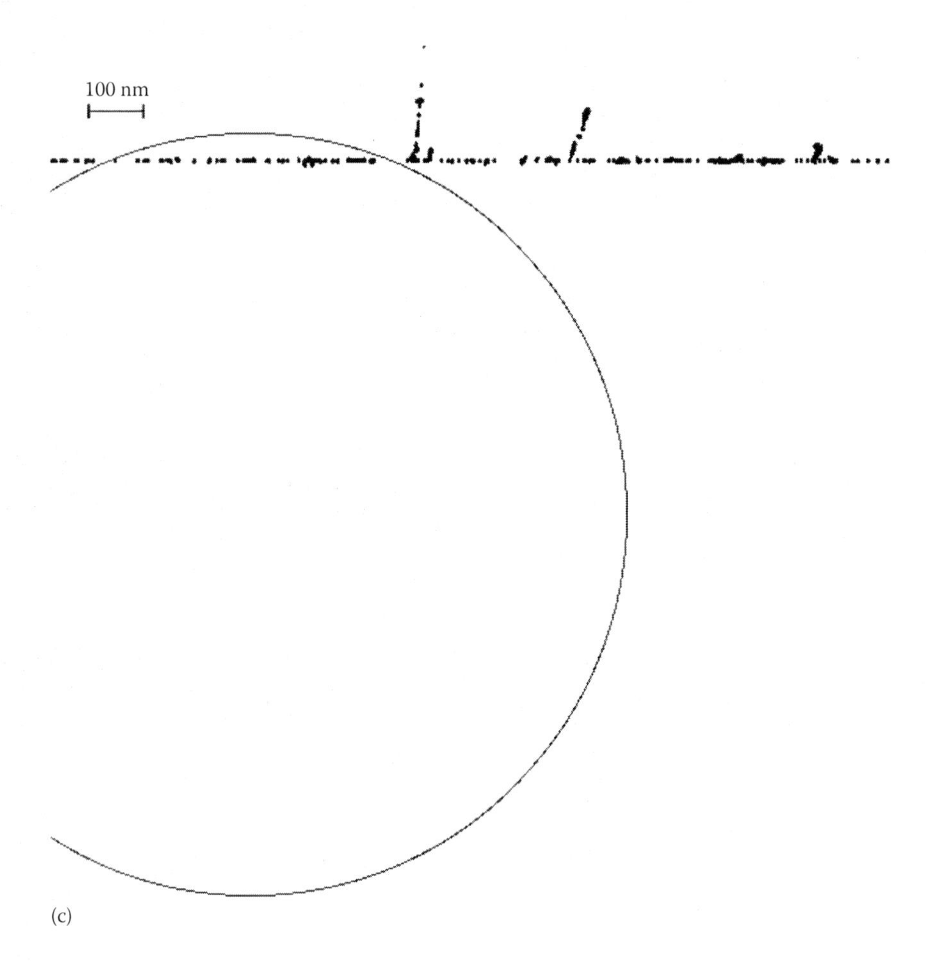

FIGURE 2.6 (CONTINUED) (c) The same particle as in (a) above but the main particle is passing only a short chord length. (*Continued*)

these smaller volumes and this track segment the fraction of volumes in which only delta electrons deposit energy becomes larger.

Figures 2.7, 2.8 and 2.9 give $\bar{y}_D$ as well as L_∞ as a function of the simulated diameter of a cylindrical unit density site with equal diameter and height for electrons, protons and carbon ions. For all three particles, a more or less strong dependence on the diameter of the volume is observed. For diameters of 100 nm or more $\bar{y}_D$ approaches L_∞. Note that for electrons, Figures 2.7, and large volumes there are still considerable differences between $\bar{y}_D$ and L_∞, and $\bar{y}_D > L_\infty$. For electrons, most protons and for carbon ions with energy larger than 60 MeV u^{-1}, $\bar{y}_D$ becomes increasingly larger than L_∞ with decreasing volume diameter, while for

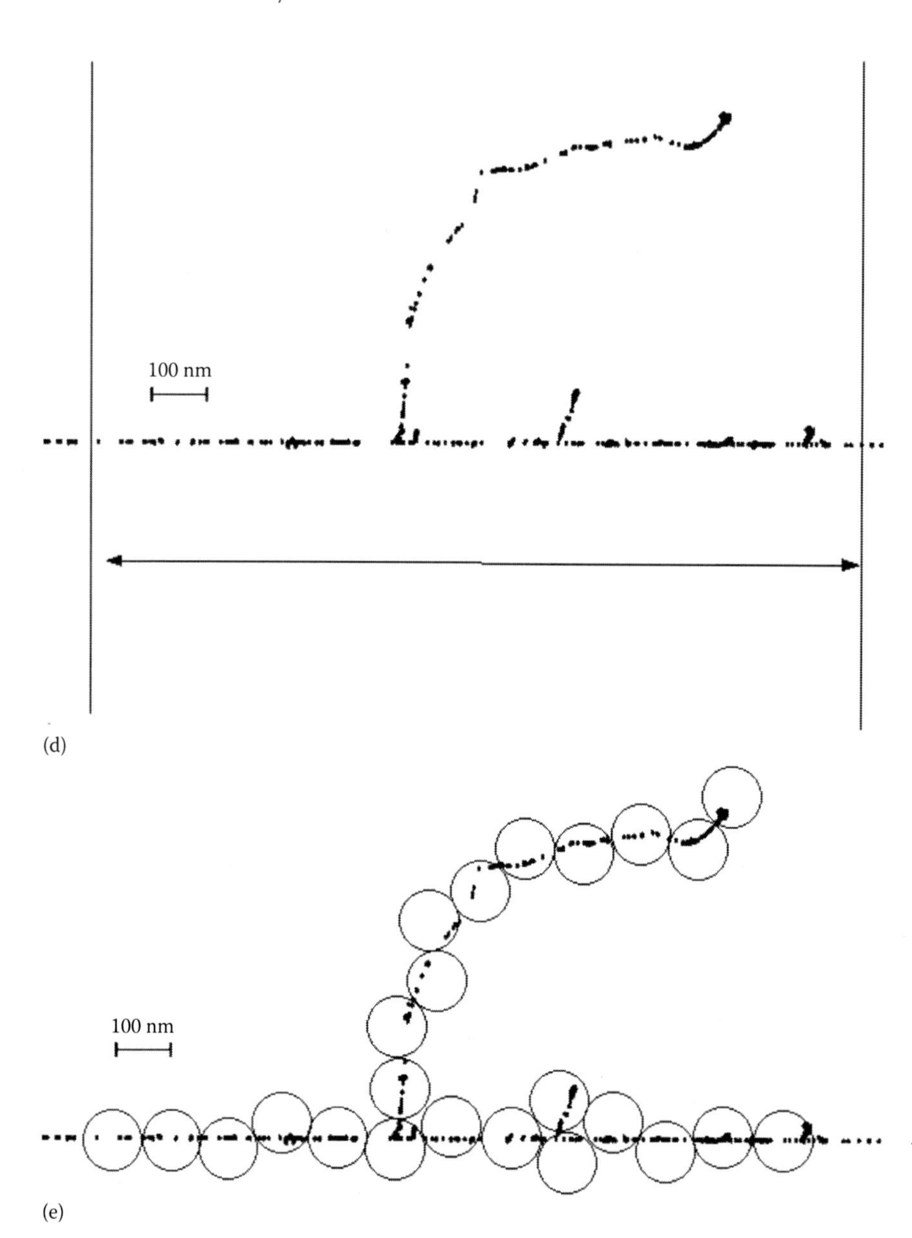

FIGURE 2.6 (CONTINUED) (d) The same track as in (a). The arrow marks the distance *dl*. When the track passes this distance it transfers the energy *dE* and $L_{\infty} = dE/dl$. The energy deposited by the main track will always be included in *dE*. (e) The same track as above. Spheres of about 100 nm in diameter cover here the track. In 11 of the 24 volumes, the energy is deposited by secondary electrons.

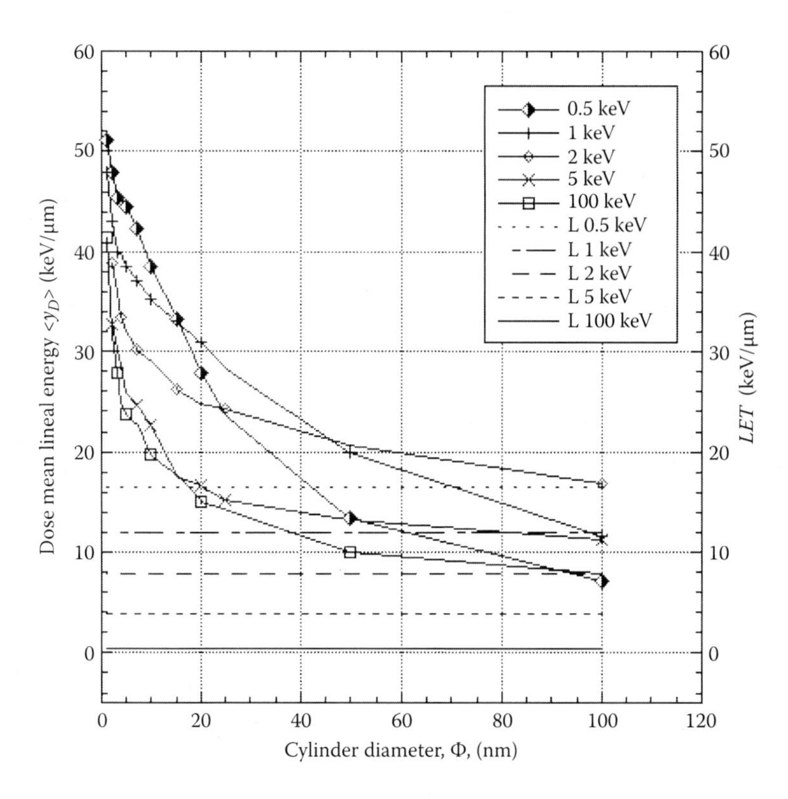

FIGURE 2.7 Comparison of dose mean lineal energy $\bar{y}_D$ and L_∞ for monoenergetic electrons with energies between 0.5 keV and 100 keV as a function of the diameter of unit density water vapour cylinders of equal diameter and height (redrawn from Lindborg et al., 2013). Horizontal lines represent L_∞ for the different electron energies. These $\bar{y}_D$ and L_∞ values originate from Nikjoo et al. (1994).

carbon ions with less than 60 MeV u^{-1}, $\bar{y}_D$ decreases with decreasing volume diameter. The reasons for these differences between $\bar{y}_D$ and L_∞ can be found by consideration of energy straggling and delta electrons. Our intention in this section is to point to a few obvious relationships seen between $\bar{y}_D$ and L_∞ and their causes. For a deeper theoretical explanation, readers are referred to other publications (Kellerer, 1985; Nikjoo et al., 2012; Rossi and Zaider, 1996).

When a charged particle experiences collisions energy is transferred and creates ionisations as well as excitations and delta electrons. The number of collisions for a given chord length varies and is given by a Poisson distribution. The expectation value of the collision number is proportional to L and the mean chord length of the volume and will be inversely

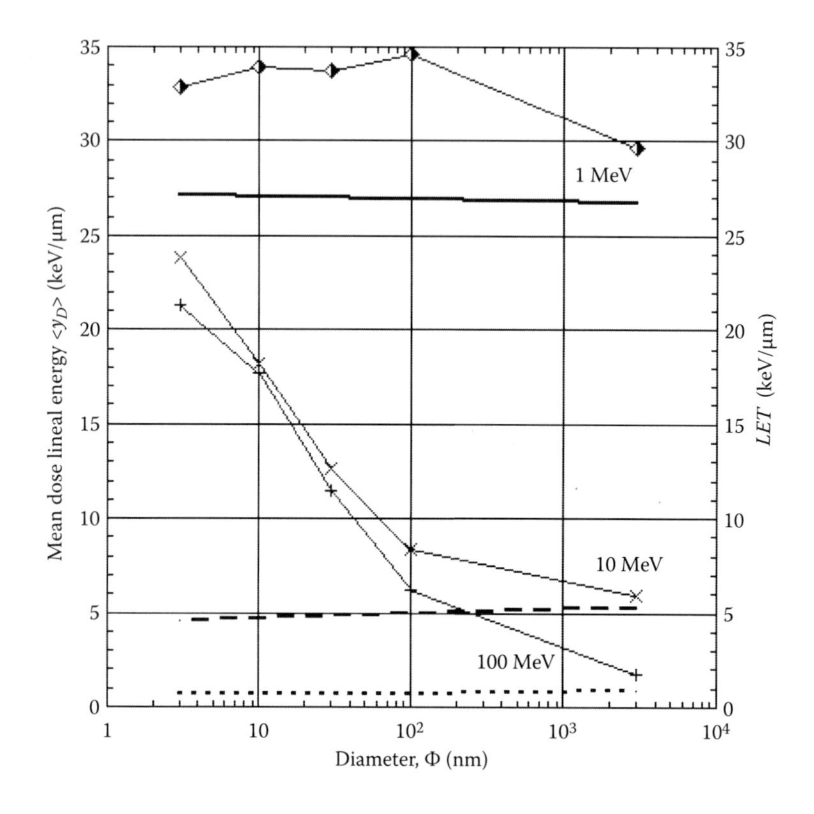

FIGURE 2.8 Comparison of dose mean lineal energy $\bar{y}_D$ and L_∞ for protons of 1, 10 and 100 MeV as a function of the diameter of unit density water vapour cylinders of equal diameter and height (redrawn from Lindborg et al., 2013). Horizontal lines represent L_∞ for the different proton energies. The $\bar{y}_D$ and L_∞ values up to 100 nm originate from Liamsuwan et al. (2010), while the values at 3 μm were published by Nikjoo et al. (2007).

proportional to the mean energy per collision. The energy transferred in a collision is also subject to fluctuation, and small energy transfers are much more likely than large energy transfers. The most important of these two distributions, number of collisions and energy transferred per collision, is the latter. The two processes constitute energy-loss straggling. It is clear that when many collisions occur extreme values of energy transferred in a single collision will be less influential and the mean lineal energy in larger site sizes will be less extreme.

A heavy charged particle track consists of a core of a few nanometres (nm) and delta electrons with ranges that may extend into the micrometre range. Therefore, with decreasing volume, delta electrons will increase in

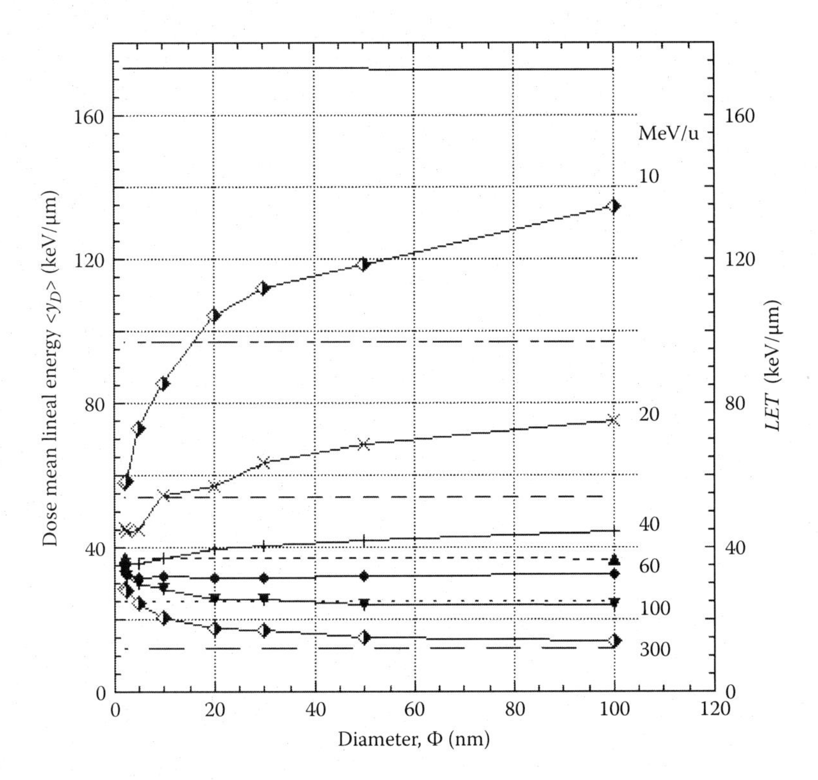

FIGURE 2.9 Comparison of dose mean lineal energy $\bar{y}_D$ and L_∞ for carbon ions of 10, 20, 40, 60, 100 and 300 MeV u^{-1} as a function of the diameter of unit density water vapour cylinders of equal diameter and height (redrawn from Lindborg et al., 2013). Horizontal lines represent L_∞ for the different carbon ions. The $\bar{y}_D$ and L_∞ values originate from Hultqvist and Nikjoo (2010).

importance (Figure 2.6d) and the low energy electrons created by a fast ion track (ion with a low LET) will have $\bar{y}_{D,e}$ larger than that of the ion itself, and as a consequence, $\bar{y}_D$ for the ion will gradually increase. For a sufficiently low energy ion track (ion with a high LET), the $\bar{y}_{D,e}$ of the delta electrons will be smaller than that of the ion track, $\bar{y}_D$, and consequently $\bar{y}_D$ will decrease with decreasing volume as gradually more volumes are only intercepted by electrons. This explanation is in line with what is observed in Figures 2.8 and 2.9.

The energy, E_e, of delta electrons scattered at angle θ can be calculated according to

$$E_e = \frac{4m_e T}{m_{hi}} \cos^2 \theta \tag{2.45}$$

where θ is the angle in which the electron is scattered, m_{hi} is the mass of the heavy ion, m_e is the mass of the electron and T is the kinetic energy of the heavy ion (see, for instance, Nikjoo et al., 2012).

Calculations show that delta electrons scattered at angles between 25° and 75° by carbon ions with energies between 20 MeV u^{-1} and 300 MeV u^{-1} are in the keV range (0.2 to 45 keV). Their penetration depths are from about 20 nm to several micrometres (Figure 2.10). The $\bar{y}_D$ of these electrons for a volume of 10 nm diameter varies approximately between 20 keV μm^{-1} and almost 40 keV μm^{-1} (Figure 2.11). These values are equal to or larger than $\bar{y}_D$ of the carbon ion itself at 300 MeV u^{-1} and

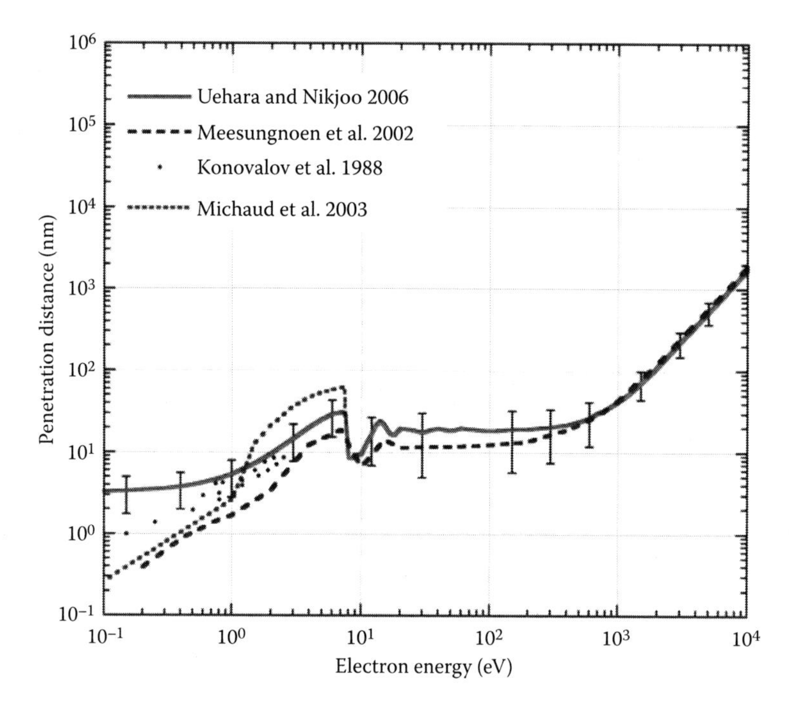

FIGURE 2.10 Variation of the electron penetration distance in water as a function of initial electron energy between 0.1 eV and 100 keV. This distance is defined as the length of the vector from the point of departure to the final position of the electron after thermalisation. Included in the figure are distances reported by several authors (Konovalov et al., 1988; Meesungnoen et al., 2002; Michaud et al., 2003; Uehara and Nikjoo, 2006) as well as the uncertainties associated with range and thermalisation distances. While the agreement in distances at 1 keV and above amongst the different authors is good, the uncertainties below 1 keV become large. The error bars show standard deviations of Monte Carlo simulation results (Nikjoo and Lindborg, 2010).

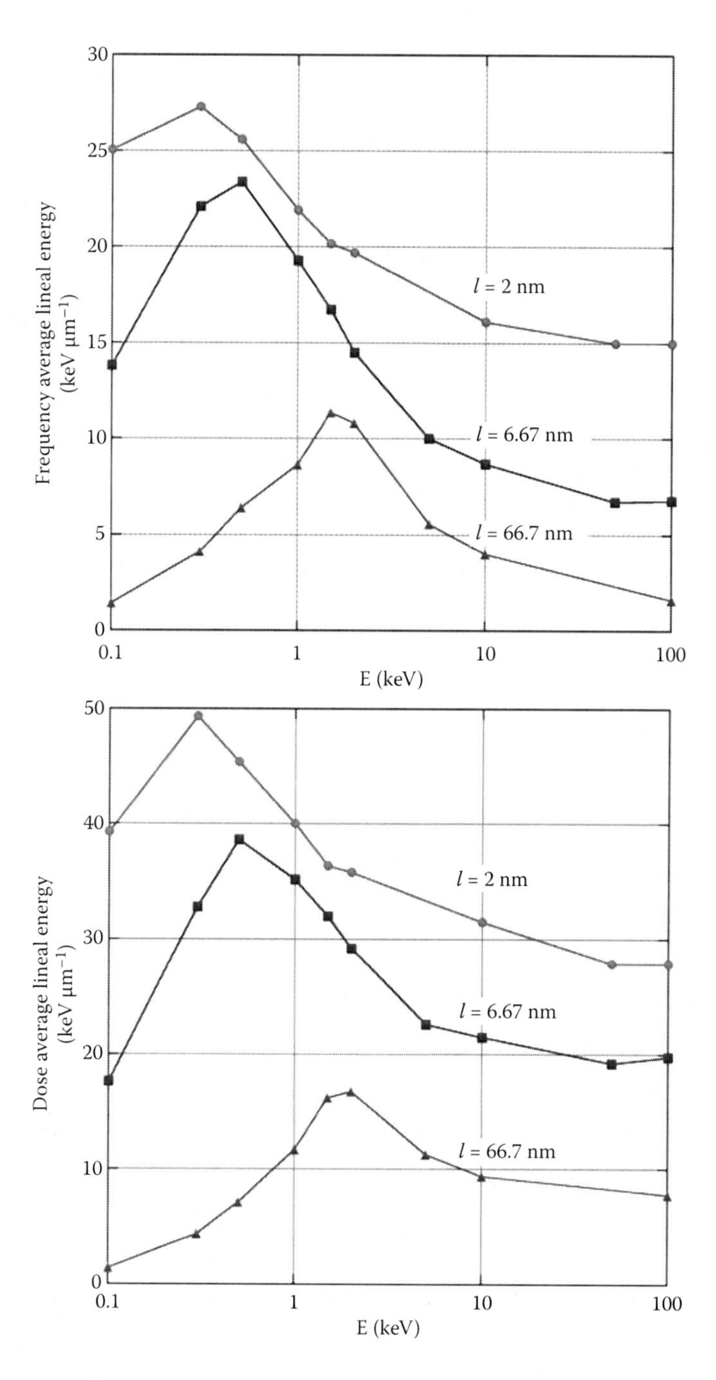

FIGURE 2.11 Frequency and dose mean lineal energies for electrons in cylindrical volumes of unit density of water vapour for volumes with three different mean chord lengths, 2 nm, 6.7 nm and 66.7 nm (Nikjoo and Lindborg, 2010).

for volumes larger than 10 nm, while for carbon ions of 20 MeV u^{-1}, the $\bar{y}_D$ of the delta electrons is smaller than $\bar{y}_D$ for the ion itself for diameters larger than 10 nm. In Figure 2.9, $\bar{y}_D$ changes more rapidly below about 30 nm for both carbon ion energies discussed. This is in accordance with the increasing amount of energy deposited by delta electrons. This suggests that delta electrons strongly contribute to the values of $\bar{y}_D$ in small-volume sites. Another factor that will increase $\bar{y}_D$ in small volumes is the mean free path between energy depositions. At 300 MeV/u, this has been reported to be about 67 nm (Hultqvist, 2011). For large volumes and for 300 MeV u^{-1} there is not much difference between the L value and $\bar{y}_D$. Here energy straggling is expected to play a minor role.

The increase observed for $\bar{y}_D$ for protons of 10 MeV and 100 MeV starts at quite large diameters (Figure 2.8) and suggests a strong influence of energy straggling. The delta electrons created by protons in this energy range will also have energies in the kiloelectron volt range and for the scattering angles 25° to 75° their energy varies between 1 keV and 200 keV. These electrons have $\bar{y}_D$ values between 20 keV µm^{-1} and 30 keV µm^{-1} at 10 nm (Figure 2.11). For protons with energies of 10 MeV and 100 MeV $\bar{y}_D$ is at 10 nm approximately 18 keV µm^{-1}. Delta electrons are therefore another likely cause for the increase of $\bar{y}_D$ with decreasing diameter. At 1 MeV $\bar{y}_D$ is not much dependent on the diameter below 100 nm. Here L is 27 keV µm^{-1} and the mean value of $\bar{y}_D$ is 33 keV µm^{-1} close to 30 keV µm^{-1} ($9L/8$). For equality between LET and lineal energy, $\bar{y}$ shall also equal L. In this case, $\bar{y}$ varies from 17 keV µm^{-1} to 23 keV µm^{-1} when going from 10 nm to 100 nm (Liamsuwan et al., 2014). Thus $\bar{y}$ is slightly lower than L (27 keV µm^{-1}). For a proton of 1 MeV the conditions for equality between L and y seem to be almost fulfilled.

For electrons the influence of energy straggling is expected to dominate. By definition, the maximum delta energy of secondary electrons is half that of the primary electron. The $\bar{y}_D$ of an electron of lower energy is usually larger than that of the primary electron, and delta electrons can be expected to contribute to the increasing $\bar{y}_D$ with decreasing volume. There is, however, an important observation. For electrons of 0.5 keV (Figure 2.7) $\bar{y}_D$ is well below L_∞ at 100 nm. Electrons of this energy have a range of only a few tens of nanometres (Figure 2.10), and the energy will be deposited in just a fraction of the volume. When this energy is divided by the mean chord length larger than the range of the particle, the lineal energy will become smaller than L_∞. Lineal energy may, under these circumstances, be less useful for characterising radiation quality.

2.3.4 Components of the Relative Variance of a Single-Event Distribution

Event distributions are quite wide, and several factors contribute to this. On condition the particle ranges are long compared to the site, the relative variance of a single-event distribution, $V_{1\mathrm{rel}}$, may be written as the sum of several relative variance components (Kellerer, 1985) and

$$V_{1\mathrm{rel}} = V_{L,\mathrm{rel}} + V_{l,\mathrm{rel}} + V_{L,\mathrm{rel}}V_{l,\mathrm{rel}} + V_{s,\mathrm{rel}} + \left(V_{F,\mathrm{rel}} + V_{m,\mathrm{rel}}\right) \tag{2.46}$$

Here
$V_{L,\mathrm{rel}}$ is the relative variance of the LET distribution.
$V_{l,\mathrm{rel}}$ is the relative variance of the track-length distribution.
$V_{s,\mathrm{rel}}$ is the relative variance of the energy straggling distribution.
$V_{F,\mathrm{rel}}$ is the relative variance of the Fano distribution.
$V_{m,\mathrm{rel}}$ is the relative variance of the gas multiplication.

The two last two terms are introduced by experimental techniques using gas ionisation devices with gas multiplication such as tissue-equivalent proportional counters. The relative variance of the LET distribution is given by

$$V_{L,\mathrm{rel}} = \frac{\bar{L}_D}{\bar{L}_T} - 1 \tag{2.47}$$

For charged particles, values of the two quantities L_T and L_D can be found in ICRU Report 16 (1970) and in Watt (1996). The relative variance of the track length distribution is given by

$$V_{\ell,\mathrm{rel}} = \frac{\bar{\ell}_D}{\bar{\ell}} - 1 \tag{2.48}$$

For a sphere with diameter d, $\bar{\ell}_D = 3d/4$, $\bar{\ell} = 2d/3$ and $V_{\ell,\mathrm{rel}} = 1/8$.

$V_{s,\mathrm{rel}}$ is dependent on the size of the volume as well as on the particles studied and is expected to dominate in small volumes.

If the same energy, ε, is repeatedly imparted to a volume the number of ions created will vary and its relative variance is given by $V_{F,\mathrm{rel}} = W/2\varepsilon$ (Kellerer, 1985). W is the mean energy needed to create an ion pair and is approximately 30 eV for propane-based tissue equivalent (TE) gas and low-LET radiation.

$V_{m,\mathrm{rel}}$ is relevant for detectors in which gas multiplication is used. In a proportional counter $V_{m,\mathrm{rel}} = \mathrm{W}/\varepsilon$. The experimental variance contributions are generally small compared to those from the other factors.

2.4 PROXIMITY FUNCTION, $T(x)$

It was realised early on that the proximity of energy deposits may play a role in understanding of the harm created by ionising radiation. Accordingly, a proximity function, $T(x)$, was defined to allow for the calculation of distances between energy deposits without necessarily having access to a Monte Carlo track structure code (Kellerer, 1985). It is also possible to experimentally determine the derivative $t(x)$ when $\bar{z}_{D,s}$ is known for spherical volumes covering a large range of diameters (Zaider et al., 1982). The function will appear in an application in Chapter 4 and is defined here.

If the transfer points of a track of specified type and energy as well as the energy deposits at those points are known, then the integral proximity function, $T(x)$, is

$$T(x) = \lim_{n \to \infty} \frac{1}{n} \sum_{j=1}^{n} \frac{\sum_i \sum_k \varepsilon_i \varepsilon_k}{\sum_i \varepsilon_i} \tag{2.49}$$

where n is the number of tracks; for any track j there are i transfer points at which energy ε_i is deposited and finally k is the number of transfer points within a distance up to x from the transfer point t_i. $T(x)$ is thus a weighted mean of the energy imparted to a spherical volume of radius x centred at an arbitrary transfer point of an arbitrary track.

The differential proximity function $t(x)$ is the derivative of $T(x)$. It is related to $\bar{z}_{D,s}$ for a randomly oriented volume or for an arbitrary volume in an isotropic field and

$$\bar{z}_{D,s} = \frac{1}{m} \int_0^\infty \frac{s(x)t(x)dx}{4\pi x^2} \tag{2.50}$$

where $s(x)\ dx$ is equal to the volume of the domain of interest that is contained in a spherical shell of radius x and thickness dx centred at a point randomly chosen in the domain of interest, and m is the mass of the domain. For a spherical volume of radius r this is

$$s(x) = 4\pi x^2 \left(1 - \frac{3x}{4r} + \frac{x^3}{16r^3}\right) \qquad x \le 2r \tag{2.51}$$

and $s(x) = 0$ for $x > 2r$. If Equation 2.51 is inserted into Equation 2.50 and $\bar{y}_D$ replaces $\bar{z}_{D,s}$ then

$$\bar{y}_D = \frac{3}{2d}\int_0^d \left(1 - \frac{3x}{4r} + \frac{x^3}{16r^3}\right) t(x)dx \tag{2.52}$$

This method was used by (Chen 2007 and 2011) to report $\bar{y}_D$ for protons of energies from 0.05 MeV to 200 MeV and for carbon ions with energies from 50 MeV to 5 GeV in volumes with diameters from a few single nanometres (nm) up to 1000 nm.

2.5 MICRODOSIMETRY DISTRIBUTIONS AND THEIR REPRESENTATION

2.5.1 Logarithmic Representation of Measured Probability Density Distributions of Lineal Energy

Microdosimetric frequency and dose distributions may be derived from single event measurements or from track structure Monte Carlo calculations. Regardless of the means of deriving the frequency and dose distributions, because of the very large range of event sizes recorded, it is usual to display the results in a logarithmic format consisting of a scale of equal logarithmic intervals of lineal energy, y. This representation of the data conveys a number of advantages, the principal one being that for a dose distribution the area under the curve defined by a specific range of lineal energy values represents the dose delivered by events in that specified lineal energy range. Furthermore, we can generalise by saying that equal areas under the distribution represent equal contributions to the absorbed dose. As an example, consider Figure 2.12, which shows a lineal energy dose distribution for a measurement in a thermal neutron beam. Events below 10 keV μm^{-1} are due to low-LET events generated by hydrogen capture gamma rays while events above 10 keV μm^{-1} are high-LET events generated by protons emitted following nitrogen capture of thermal neutrons. The roughly equal area of the distribution up to 10 keV μm^{-1} with that beyond 10 keV μm^{-1} indicates that when thermal neutrons interact with tissue, the absorbed dose is approximately equally split between the interaction of gamma rays and protons. This useful characteristic of the logarithmic representation of microdosimetric data can be expressed mathematically in the following manner.

If in Figure 2.12 we take an infinitesimal small logarithmic interval of lineal energy $d\ln y$ and multiply this by its height $yd(y)$ we obtain the area under that region of the curve, $yd(y)d\ln y$. As the logarithmic

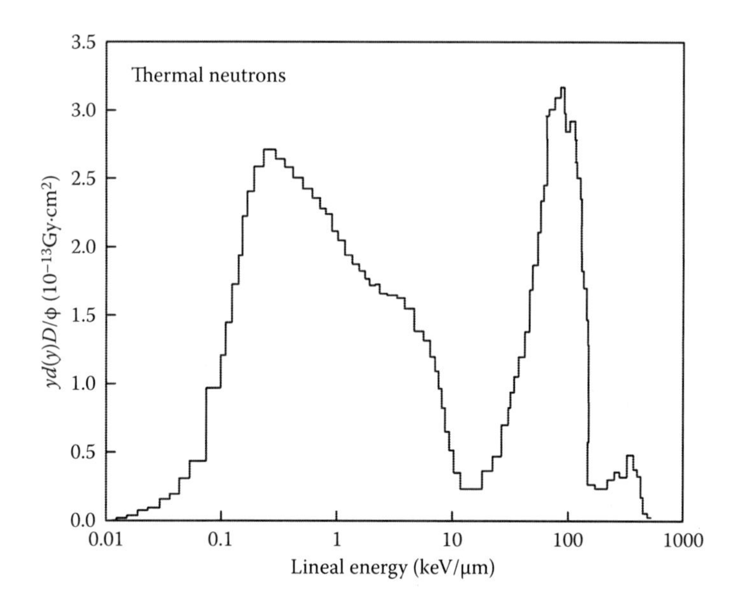

FIGURE 2.12 Generic single-event spectrum measured with a tissue-equivalent proportional chamber (TEPC) in a thermal neutron beam in front of a tissue-equivalent phantom. The *yd(y)* axis, in this case, is given in terms of absorbed dose per unit fluence of thermal neutrons.

differential $d\ln y = \frac{1}{y}dy$, where dy is the linear differential, the product becomes

$$yd(y)d\ln y = \frac{yd(y)dy}{y} = d(y)dy$$

Thus

$$\int_{y_1}^{y_2} d(y)dy = \int_{y_1}^{y_2} yd(y)d\ln y$$

Further details of the presentation of experimental frequency and dose data in a logarithmic format are covered in Chapter 3, which discusses experimental methods and procedures.

2.5.2 The Additivity of Dose Fractions

A lineal energy dose distribution may be divided into arbitrary lineal energy intervals and the fractions of the energy imparted, and thereby the absorbed dose in those intervals is given by the integrals of the lineal energy within those intervals. Thus

$$\bar{y}_D = \int_{y_{\min}}^{y_{\max}} yd(y)dy = \int_{y_{\min}}^{a} yd(y)dy + \int_{a}^{b} yd(y)dy + \int_{b}^{y_{\max}} yd(y)dy \tag{2.53}$$

or

$$\bar{y}_D = d_1\bar{y}_{D,1} + d_2\bar{y}_{D,2} + d_3\bar{y}_{D,3} \tag{2.54}$$

where d_1, d_2 and d_3 are the dose fractions for the lineal energy intervals 1, 2 and 3.

This possibility is of interest if different weighting factors are needed, for instance, to estimate the relative biological effectiveness (RBE) or the dose equivalent in radiation protection. After measurements in a radiation environment where neutrons are present, it is usually possible to distinguish different dose components in the dose distribution: d_1 could correspond to the dose fraction from low-LET radiations, d_2 to the dose fraction from protons and d_3 that from high-LET heavy ions.

This technique can also be used in calculations of $\bar{y}_D$ at any specific point in a radiation field. It requires that the dose fraction for each of the different particles, d_i, as well as their energy distributions, $\Phi_i(E)$, and mass stopping power, $\left(\frac{S_i(E)}{\rho}\right)$, at a specific point in a radiation field are known. The dose fraction of particle i is then given by

$$d_i = \frac{\int_0^\infty \Phi_i(E)\left(\frac{S_i(E)}{\rho}\right)dE}{\sum_i \int_0^\infty \Phi_i(E)\left(\frac{S_i(E)}{\rho}\right)dE} \tag{2.55}$$

If for each particle, i, and energy, E, $\bar{y}_{D,i}(E)$ is known, $\bar{y}_{D,i}$ at this point is

$$\bar{y}_{D,i} = \frac{\int_0^\infty \Phi_i(E)\left(\frac{S_i(E)}{\rho}\right)\bar{y}_{D,i}(E)dE}{\int_0^\infty \Phi_i(E)\left(\frac{S_i(E)}{\rho}\right)dE} \tag{2.56}$$

The $\bar{y}_D$ value for the radiation beam is obtained by weighting the relative dose fraction with the $\bar{y}_{D,i}$ values of the primary and secondary ions:

$$\bar{y}_D = \int yd(y)dy = \sum d_i \bar{y}_{D,i} \tag{2.57}$$

This technique has been used by Liamsuwan et al. (2014), Hultqvist et al. (2010) and Lillhök et al. (2007a) to calculate $\bar{y}_D$ in volumes in the nanometre range. Tables of both $\bar{y}(E)$ and $\bar{y}_D(E)$ for several particles have been produced by Nikjoo and colleagues (Nikjoo et al., 2011).

2.5.3 *F*(*z*,*D*) Multiple-Event Frequency Distribution

Once the single-event frequency distribution of the specific energy is known, it is possible to calculate the multi-event frequency distribution. The number of events, ν, for a given absorbed dose, *D*, is given by the Poisson distribution (Equation 2.7). The specific energy per event varies according to the distribution $f_s(z)$ and the multi-event distribution is

$$f(z,D) = \sum_{\nu=0}^{\infty} e^{-\bar{n}} \frac{\bar{n}^{\nu}}{\nu!} f_{\nu}(z) \tag{2.58}$$

where $\bar{n}$ and ν have the meanings defined previously and ν = 2, 3…. and $f_\nu(z)$ is the ν-fold convolution of the single-event spectrum $f_s(z)$. A computer code for calculating this has been published by Kellerer (1985).

Calculation of multiple-event distributions using Fourier transforms has been suggested by Roesch (1977). A simplified version of the method has been published by Santa Cruz et al. (2001a,b). An alternative approach to calculate the spread of the multi-event distribution was published by Villegas et al. (2013) and Villegas and Ahnesjö (2016). The main application seems to have been for internal dosimetry of heavy particles such as alpha emitters (Palm et al., 2004). Experimentally determined multi-event distributions will be dealt with in Chapter 3.

2.6 SUMMARY

Energy deposit, ε_i, in a single interaction is the most fundamental quantity in dosimetry.

Energy imparted, ε, is the sum of all energy deposits in a specified volume and is defined for both single events, denoted ε_s, and multi-events, denoted ε.

Lineal energy, *y*, is $y = \varepsilon_s/\bar{\ell}$, where $\bar{\ell}$ is the mean chord length of the volume. This quantity, *y*, is defined only for single events.

TABLE 2.2 Definitions of Expectation Values for Lineal Energy and Specific Energy Distributions

Expectation Values of Single-Event Distributions		Expectation Value of a Multi-Event Distribution
$\bar{y} = \int_{ymin0}^{ymax} yf(y)dy$	$\bar{y}_D = \frac{1}{\bar{y}} \int_{ymin}^{ymax} y^2 f(y)dy$	
$\bar{z}_s = \int_{z_{min}}^{z_{max}} z_s f_s(z)dz$	$\bar{z}_{D,s} = \int_{z_{min}}^{z_{max}} z_s^2 f_s(z)dz$	$\bar{z} = \int_0^{\infty} zf(z)dz$

Specific energy z, is $z = {}^{\varepsilon}/_{m}$, where m is the mass of the volume and ε is due to either a single event or multi-events.

The quantities cited in the preceding text are stochastic quantities and are described by probability distribution functions from which their expectation values can be determined (Table 2.2). The mean values $\bar{\varepsilon}, \bar{y}, \bar{z}$ are all non-stochastic quantities. When a quantity represents a single event it has index s with the exception of y, which is defined only for single events and therefore there is no need for extra notation.

Specific energy, z_s (Gy), and lineal energy, y (keV μm^{-1}) are related and

$$z_s = \frac{\varepsilon_s}{m} = \frac{y\bar{\ell}}{\varrho V} = ky \tag{2.59}$$

where V (μm^3) is the volume of the microscopic site with mean chord length $\bar{\ell}$ (μm) and density ϱ (g cm^{-3}). When $\varrho = 1$ g cm^{-3}, k is

$$k = 0.1602\bar{\ell}/V\ \left(\text{Gy}/\left(\text{keV}\,\mu\text{m}^{-1}\right)\right) \tag{2.60}$$

For any convex volume with surface area S (μm^2),

$$\bar{\ell} = \frac{4V}{S} \tag{2.61}$$

For a sphere with diameter d (μm),

$$k = 0.204/d^2 \tag{2.62}$$

The number of events n creating the absorbed dose, D, is a stochastic quantity ruled by Poisson statistics. If the single-event mean specific energy is $\bar{z}_s$ and $\bar{n}$ the mean number of events, then $D = \bar{n}\bar{z}_s$.

If the probability for zero-events is negligible, $\bar{z} = D$ and the relative variance, V_r, of the multi-event frequency distribution is $V_r = \bar{z}_{D,s}/D$, where $\bar{z}_{D,s}$ is the single-event dose-mean specific energy.

LET is $L_\infty = dE/dl$, where dE is the energy transferred to matter when a charged particle passes the distance dl. LET is a non-stochastic quantity. L_∞ always includes the energy lost by the primary charged particle and its delta electrons.

The quantity y is the appropriate one for characterising radiation quality of charged particles, in particular for volumes with dimensions of less than about 100 nm. The randomly positioned volumes along a particle track will sometimes include the primary charged particle and sometimes not. With decreasing volume, delta electrons will increase in importance for the value of y.

CHAPTER 3

Experimental Microdosimetry

3.1 INTRODUCTION

In experimental microdosimetry two independent measurement methods have dominated: *pulse height analysis* (PHA) of single events and *variance–covariance* (VC) analysis of multi-events. In the former, the energy imparted by single events is analysed and by recording a large number of single events a *single event distribution* is created. From this distribution the various microdosimetric averages discussed in the previous chapter can be obtained.

In the VC method the energy imparted by multi-events is analysed. This energy is determined in a series of measurements, which are equally long in time. The mean of the created multi-event distribution, $\bar{\varepsilon}$, times its relative variance is the dose mean energy imparted by single events, $\bar{\varepsilon}_{D,s}$. The technique is based on measurements either with electrometers or with charge-sensitive pulse amplifiers. The integrations have to be long enough to include always at least one energy deposition event.

Both measurement techniques are today well established and have found several applications, particularly in radiation fields with dose components from different charged particles such as onboard aircraft, in space applications and in the nuclear industry (Chapter 5). Microdosimetric measurements are also recommended for quantifying radiation quality in radiation therapy when beams of protons, ions and neutrons are being used (Chapter 4). Recently the International Commission on

Radiation Units & Measurements (ICRU) has also recommended that in cases in which the energy distribution of particle irradiance cannot be deduced or measured, lineal energy and energy deposition event rates may be the best way in which to quantify low-dose and other heterogeneous exposures (ICRU, 2011).

To date proportional counters have been the most frequently used detector for experimental microdosimetry, but ionisation chambers, solid-state detectors and more recently gas electron multiplier (GEM) detectors have also been investigated for their use in microdosimetric measurements. Reviews of the detector development have been presented by, for instance, Braby (2015), EURADOS (1995), Kliauga (1990a) and Schuhmacher and Dangendorf (2002).

In most applications, a mean chord length in the range 0.1 μm to 2 μm is used. However, measurements with the PHA method have been reported down to mean chord lengths of about 50 nm and with the VC method down to 6 nm. Results below 50 nm become more uncertain. Volumes in the nanometre range have gained an increasing interest, and other techniques have been introduced or are under development and are briefly covered in Section 3.6 of this chapter.

Today calculations of microdosimetric distributions can be made with several condensed history Monte Carlo codes as well as with track-structure Monte Carlo codes. Results for volumes down to 1 nm can be found for both low- and high-linear energy transfer (LET) radiations. It also should be mentioned that typically not all condensed history codes, for example, MCNPX, FLUKA and PHITS, give the same results when used to model the same radiation fields and detector responses; therefore care must be taken to assess properly the performance of the code to be used and wherever possible carry out comparisons between modelling results and benchmark measurements (Ali et al., 2014).

3.2 SIMULATION OF TISSUE

If a charged particle interacts with the same number of atoms in a gas as it would in tissue and if the atoms are the same in both gas and tissue, then to a good approximation any effect of the difference in density can be ignored for the volumes most frequently simulated. As the density ratio for tissue to gas is roughly 1000, a tissue volume of 1 μm in diameter can then be represented by a gas volume roughly 1000 times larger or about 1 mm in diameter at normal pressure. The dimension of the gas volume can be increased and still simulate the same tissue volume, if the density of the gas is reduced. In this way a gas volume large enough to be used as a detector can be created if it is equipped with electrodes and connections for establishing an electric field for charge collection. The fundamental requirement is that the gas filling and the wall of the counter surrounding the cavity are composed of the same atoms in the same proportion as those found in tissue. Harald H. Rossi and co-workers were the first to build a proportional counter that fulfilled these requirements and in so doing initiated the scientific field of experimental microdosimetry (Rossi and Rosenzweig, 1955a). Proportional counters of this type have come to be known as tissue-equivalent proportional counters (TEPCs).

3.2.1 Relationships for the Simulation of a Soft Tissue Site by a Gas

The tissue site (subscript t) and the detector gas cavity (subscript det) are characterised by their density (ϱ) and mass stopping power ($dE/\varrho dx$). The volumes are characterised by their mean chord lengths $\bar{\ell}_t$ and $\bar{\ell}_{\text{det}}$, where $\bar{\ell}_{\text{det}}$ is defined by the physical dimensions of the detector.

The mean energy imparted by a charged particle crossing the two volumes along their respective mean chord lengths is

$$\bar{\varepsilon}_{t,s} = \left[\frac{dE}{\varrho dx}\right]_t \varrho_t \bar{\ell}_t \tag{3.1}$$

$$\bar{\varepsilon}_{\text{det},s} = \left[\frac{dE}{\varrho dx}\right]_{\text{det}} \varrho_{\text{det}} \bar{\ell}_{\text{det}} \tag{3.2}$$

If the atomic compositions of the tissue and the detector gas are identical and therefore their mass stopping powers the same, and if the density of the detector gas is adjusted so that

$$\varrho_{\text{det}} \bar{\ell}_{\text{det},} = \varrho_t \bar{\ell}_t \tag{3.3}$$

then a charged particle will deposit the same mean energy when crossing the mean chord lengths in both the tissue volume and the gas volume as

$$\bar{\varepsilon}_{t,s} = \left(\frac{dE}{\varrho dx}\right)_t \varrho_t \bar{\ell}_t = \left(\frac{dE}{\varrho dx}\right)_{\text{det}} \varrho_{\text{det}} \bar{\ell}_{\text{det}} = \bar{\varepsilon}_{\text{det},s} \tag{3.4}$$

The mean lineal energy, $\bar{y}$, and mean specific energy of single events, $\bar{z}_s$, are easily derived by dividing $\bar{\varepsilon}_{t,s}$ by $\bar{\ell}_t$ and m_t, respectively, m_t being the mass of the simulated tissue volume with density ϱ_t and mean chord length

$\bar{\ell}_t$. This is the basic relation for simulation of a small tissue volume with a gas volume.

If r_t is the radius of a tissue sphere and r_{det} the radius of a spherical detector, it follows from Equation 3.3

$$\varrho_{det} = \varrho_t \frac{r_t}{r_{det}} \tag{3.5}$$

Similarly the cross-sectional areas A_t and A_{det}, the volumes V_t and V_{det} and the masses m_t and m_{det} are related as follows:

$$A_{det} = A_t \frac{r_{det}^2}{r_t^2} \tag{3.6}$$

$$V_{det} = V_t \frac{r_{det}^3}{r_t^3} \tag{3.7}$$

$$m_{det} = \varrho_{det} V_{det} = \varrho_t \frac{r_t}{r_{det}} V_t \frac{r_{det}^3}{r_t^3} = \varrho_t V_t \frac{r_{det}^2}{r_t^2} = m_t \frac{r_{det}^2}{r_t^2} \tag{3.8}$$

If the detector is housed in a vacuum tight container the gas density may be reduced by lowering the gas pressure and an arbitrarily small tissue-equivalent (TE) mass can be simulated. If P_0 and $\varrho_{det,0}$ are the gas density at normal pressure, P_0, and 20°C and P_x and $\varrho_{det,x}$ are the corresponding values at pressure x and 20°C then

$$\frac{P_x}{P_0} = \frac{\varrho_{det,x}}{\varrho_{det,0}} \tag{3.9}$$

From Equations 3.3 and 3.9 it follows

$$\varrho_{det,x} \bar{\ell}_{det} = \frac{P_x}{P_0} \varrho_{det,0} \bar{\ell}_{det} = \varrho_t \bar{\ell}_t \tag{3.10}$$

or

$$\frac{P_x}{P_0} = \frac{\varrho_t \bar{\ell}_t}{\varrho_{det,0} \bar{\ell}_{det}} \tag{3.11}$$

For example, we can calculate by how much the gas pressure must be reduced in a spherical detector with mean chord length $\bar{\ell}_{det}$= 10 mm, filled with TE gas of density ϱ = 1 kg m^{-3} at 100 kPa and 20°C, to simulate a tissue volume with mean chord length 1 μm and density $\varrho_t = 10^3$ kg m^{-3}. From Equation 3.11 it follows: $\frac{P_x}{P_0} = 0.1$. By reducing the gas pressure to

1/10 of normal atmospheric pressure the 10-mm mean chord length of the detector gas volume will simulate a tissue volume with a mean chord length of 1 μm. For a precise calculation of the required pressure an accurate value of the gas density is obviously needed (see Table 3.2).

When many particles are crossing the detector simultaneously (multi-event situation), the simulation principle will remain. This is so because each *single* particle, *i*, will deposit energy, $\varepsilon_{t,i}$, along a chord for which the simulation principle is true, $\varepsilon_{t,i} = \varepsilon_{det,i}$. However, for a radiation field with a given particle fluence, the number of particles crossing the detector volume will be much larger than the number of particles crossing the tissue volume of unit density. Consequently, the mean energy imparted by multi-events in the detector, $\bar{\varepsilon}_{det}$, has to be reduced by a factor proportional to the ratio of the cross-sectional areas of the two volumes, Equation 3.6, and

$$\bar{\varepsilon}_t = \bar{\varepsilon}_{det}\frac{A_t}{A_{det}} \quad \text{which for a sphere becomes} \quad \bar{\varepsilon}_t = \bar{\varepsilon}_{det}\frac{r_t^2}{r_{det}^2} \tag{3.12}$$

If we divide $\bar{\varepsilon}_t$ by m_t and $\bar{\varepsilon}_{det}$ by m_{det} it follows from Equations 3.8 and 3.12

$$\bar{z}_{det} = \frac{\bar{\varepsilon}_{det}}{m_{det}} = \frac{\bar{\varepsilon}_t \frac{r_{det}^2}{r_t^2}}{m_t \frac{r_{det}^2}{r_t^2}} = \frac{\bar{\varepsilon}_t}{m_t} = \bar{z}_t \tag{3.13}$$

Thus the mean specific energy imparted by multiple events in a detector is the same as the mean specific energy in a tissue volume of density 1000 kg m^{-3} on condition the fluence of particles is the same. As $\bar{z}_{det} = D$, it follows that $\bar{z}_t = D$, which we already know from Chapter 2.

3.2.2 Tissue-Equivalent Matter

Most detectors are made of a plastic named A-150. This plastic was developed at the Benedictine College, Kansas, US by Dr Shonka (Shonka et al., 1958) and has been available either as granulates or as blocks and is now available through Standard Imaging. Table 3.1 shows the elemental composition of A-150; its density has been determined to be 1120 kg m^{-3} (Smathers et al., 1977). The table shows that the high content of oxygen in tissue has to a large extent been replaced by carbon. This provides the plastic with sufficient conductivity for it to be used in the construction of ion chambers and proportional counters. The photon mass absorption coefficients and the neutron kerma factors for carbon and oxygen are

TABLE 3.1 Elemental Composition in Percent by Weight of ICRU Tissue, A-150 and TE Gases

	H	C	N	O	F	Ca	Reference
ICRU tissue	10.1	11.1	2.6	76.2			ICRU (1989)
A-150	10.2	76.8	3.6	5.9	1.7	1.8	ICRU (1983)
Methane-based TE gas	10.2	45.6	3.5	40.7			ICRU (1983)
Propane-based TE gas	10.3	56.9	3.5	29.3			ICRU (1983)
Propane	18.2	81.8					

similar at the energies that are most often encountered in radiation protection, and the exchange of oxygen with carbon in the A-150 composition does not significantly disturb the tissue equivalence of the material. Smathers et al. (1977) have collected data from several reports on the composition of the plastic and recommended an elemental composition shown in Table 3.1.

A major component of A-150 is nylon, which is hygroscopic, and the content of water vapour may increase the weight of the A-150 by up to about 10%. It is therefore important to outgas the detector before it is finally filled with a TE gas. When sealed detectors are used, changes in gas multiplication have been observed as a result of changes in temperature (Kunz et al., 1992; Melinder, 1999). A decrease of 15% was reported by Melinder when the temperature increased from 22°C to 34°C with fresh methane and fresh propane-based TE gas in a large spherical TEPC. One explanation suggested is that water molecules adsorbed in the wall enter the gas when the temperature increases. If the detector is evacuated at an elevated temperature, out-gassing occurs more rapidly (Section 3.7). The effect of heating pieces of A-150 up to 120°C for 100 h has been investigated and the impact on the content of hydrogen, carbon and nitrogen was found to be insignificant when samples were chemically analysed before and after heating (Lindborg, 1974). Kliauga (1990a) has reported a 5% shrinkage of the lineal dimension of A-150 with age in particular for a wall-less detector. Waste material retrieved from machining A-150 can be recycled without significant changes in electrical properties (Waker, 1988). Minor changes in the A-150 elemental composition have been reported anecdotally among counter constructors to have occurred and it may be necessary to investigate the consequences of this in high-accuracy measurements for radiotherapy or in very energetic radiation fields encountered in space or around high-energy particle accelerators.

TABLE 3.2 Partial Pressure Components of the TE gases as well as the Density, $\varrho_{t,0}$, at 20°C and 100 kPa

Gas	Components in Percent by Partial Pressure				ϱ (kg m^{-3})	Pressure at 1 μm in a Detector with 10 mm Diameter (kPa)
	CH_4	C_3H_8	CO_2	N_2		
Methane-based TE gas	64.4	0	32.5	3.1	1.05	9.524
Propane-based TE gas	0	55	39.6	5.4	1.798	5.562
Propane		100			1.85	5.396

Source: ICRU (International Commission on Radiation Units & Measurements). Report 36: Microdosimetry. Bethesda, MD: ICRU (1983).

A few reports on alpha emitting impurities in A-150 have been published. A small number of unexpected large events was reported by Menzel et al. (1989) and it was suggested that these events were related to alpha-emitting impurities in A-150. From measurements in a low-level radiation background laboratory situated in a salt mine 490 m deep an unexpected dose rate of 10 nGy h^{-1} of high-LET events (or 13 to 15 events h^{-1}) was reported with a cylindrical TEPC with 10 cm height and 10 cm diameter. Measurements with poly allyl diglycol carbonate (PADC or CR-39) film attached to a piece of A-150 plastic revealed an alpha track density of 0.0159 + 0.0013 cm^{-2} h^{-1} corresponding to seven events per hour in the aforementioned detector (Lillhök, 2007b). Obviously this impurity affects only low-level radiation protection measurements with large detectors.

The three most used counter-fill gases are methane-based TE gas, propane-based TE gas and pure propane. They are defined in Tables 3.1 and 3.2. The last column in Table 3.2 gives the pressure corresponding to 1 μm of tissue in a sphere with 1 cm diameter at 20°C. Mass stopping power tables for A-150 as well as TE gas can be found in ICRU Reports 36 (1983), 49 (2007), Watt (1996) and www.SRIM.org, www.nist.gov/pml/data/radiation.cfm.

3.3 DETECTORS

Several types of detectors have been used in experimental microdosimetry. For single-event measurements tissue equivalent proportional counters have been the dominant instrument used but more recently GEM detectors have also been investigated for their potential in microdosimetry because of their small size, relatively simple construction and the possibility of particle tracking (Farahmand et al., 2004). Solid-state detectors are now also being intensively studied for applications in radiation protection, as they are very small, rugged, need no vacuum containment and have low power consumption (Agosteo and Pola, 2011). The tissue volume they may simulate is defined by the depletion layer and is typically a few micrometres. The main detection material in solid-state devices is silicon, which is not tissue equivalent, and so some corrections to the measured data have to be applied as well as hydrogenous converters for neutron microdosimetry. In future years it is expected that these newer devices will become much more widely used in experimental microdosimetry although the traditional TEPC is likely to remain the 'standard' device against which the performance of other detectors is measured. For multi-event measurements where the signal is larger and which may cancel the need for gas multiplication ionisation chambers have been extensively used.

Originally detectors for experimental microdosimetry were developed to simulate tissue equivalent volumes with diameters of about one or a few micrometres and were used to characterise radiation fields used in radiobiology experiments or as survey meters in mixed radiation fields. Later, interest developed in applying the methods to both personnel dosimetry as well as radiation therapy. In the first case, a detector of small weight and size is required while in radiation therapy a detector that can operate at high dose rates is needed. An advantage for both these applications are detectors of small size as well as compact electronics, and this is still proving to be a challenge for the widespread use of microdosimetric methods. Ultra-miniature detectors to simulate smaller volumes (volumes below 100 nm) and for operation in higher dose-rate fields have been developed and reported by various research workers. Commercially produced instruments are available (Far West Technology).

3.3.1 Proportional Counters

Proportional counters are the most established detectors for measurement in the field of experimental microdosimetry. Figure 3.1 shows a typical spherical proportional detector developed at Columbia University. Counter diameters are usually between 10 mm and 150 mm although some smaller counters have been developed. With these typical sizes, the pressure needed to simulate 1 μm is in the range of 9.5 kPa to 0.9 kPa (Table 3.2). At very low pressures, P, and high electric field, E, the reduced electric field (E/P) in the counter is such that the gas avalanche is no longer confined to just a very narrow volume of a few anode-wire diameters around the central electrode and will spread farther out in the gas cavity

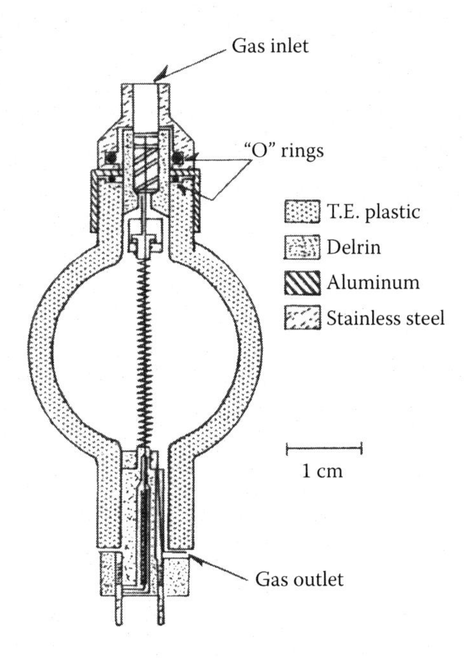

FIGURE 3.1 A solid wall TEPC made at Columbia University. The spiral around the centre wire is called the helix and is held at a positive potential with respect to the wall. The centre wire is at a positive potential with respect to the helix. The potential difference between the helix and the centre wire is much larger than between the helix and the wall. With this arrangement, the gas gain avalanche will be contained within the helix. (From ICRU (International Commission on Radiation Units & Measurements). Report 36: Microdosimetry. Bethesda, MD: ICRU (1983).)

away from the anode electrode, and as a consequence the gas multiplication will become dependent on the position of the ionisation within the counting gas. If a charged particle creates an ionisation inside the avalanche volume, it will not experience the full gas multiplication and the resolution of the proportional counter will deteriorate. This is the main reason why there is a limit to the lowest site sizes that can be simulated by a proportional counter to around a few tenths of a micrometre (Ségur et al., 1995).

3.3.1.1 Design

The volume in which the gas multiplication occurs depends on the design of the counter, in particular the diameters of the gas cavity and the central electrode as well as the TE gas pressure. The homogeneity of the gas multiplication along the central electrode is also dependent on the shape of the detector. In a cylindrical detector, because of the uniformity of the electric field along the anode wire it is easier to obtain uniformity of the gas gain than it is in a spherical detector, and the sensitive volume of the detector can be defined by the use of field tubes (Waker, 1982). To overcome the non-uniformity of the electric field and hence gas gain in spherical counters a number of arrangements have been considered, perhaps the most well known being the use of a helix (a spiralling wire surrounding and in close proximity to the central electrode) which, in effect, converts the spherical counter into a cylindrical proportional counter that is gas coupled to a spherical charge collecting volume. TE-proportional counters using this technique have been given the special name "Rossi counters" named after Harold Rossi, who along with Rosenzweig first designed such counters (Rossi and Rosenzweig, 1955). The potential of the helix wire is typically 80% of the potential at the central electrode and has the effect that the strong electric field is limited to the inside of the helix, thereby forcing the gas multiplication to stay inside this volume. A difficulty resulting from the use of a helix is that it may be sensitive to detector movements or vibrations causing microphony that can be a serious source of electronic noise limiting the lowest event sizes that can be measured. Another complicating factor can be the transparency of the helix and that some of the electrons created between the counter wall and the helix may get collected by the helix instead of the anode wire.

Another technique to make the electric field inside a sphere uniform is to weaken the electric field at the poles of the sphere by effectively increasing the diameter of the collecting electrode and increasing the distance between the central wire and the spherical wall. Such a design, known as 'single-wire counters', was first proposed by Benjamin et al. (1968) for spherical detectors used as proton-recoil neutron spectrometers and adopted by Far West Technology for TEPC design. The efficacy of this method of producing field and gas-gain uniformity has been experimentally investigated by Waker (1986) and shown to provide gas gain uniformity to within 1% of the entire counting volume.

Another technique to avoid the problem with the inhomogeneous electric field in a spherical detector has been presented by Perez-Nunez and Braby (2011). Here the spherical shell is cut into slices and insulating material inserted between the slices. Different potentials were then applied to the different parts of the shell. A similar solution was adopted for a large-volume cylindrical counter where the use of traditional field tubes would have led to a large wasted insensitive volume. In this case, each end plate of the cylindrical cathode was divided into concentric circles separated by a thin insulating layer and each ring given the correct electric potential for its distance from the anode (Verma and Waker, 1992).

A design option for field uniformity for cylindrical counters other than the use of field tubes, which significantly reduce the size of the sensitive volume of the detector, is to make a cylindrical counter so long that the volumes at the ends in which the multiplication is changing become small in comparison with the total volume (Gerlach et al., 2002). However, Waker has shown that the non-uniformity of the field even in an elongated counter can affect the counter gas gain farther away from the counter ends to a significant degree. Consequently an elongated counter was constructed that employed the same principle used in 'single-wire' counters of increasing the anode diameter at the ends of the counter and moving the cathode material farther away. Using this method a field uniformity of +2.5% was achieved over 62% of the anode length and less than 7% for the entire anode length for a cylindrical counter with length to diameter of 10:1 (Waker et al., 2011).

Originally detectors were made spherical, thereby making them in principle directionally independent. However, cylindrical detectors are often easier to construct, especially in small research workshops without moulding facilities, and such detectors are frequently used in experimental microdosimetry. The probability that a particle is travelling along a chord length between l and $l + dl$ of the counter is $t(l)d(l)$, where $t(l)$ is the track length probability distribution. In a sphere this distribution is triangular. The influence of the track length distribution on the measured distribution was described in Chapter 2 (Section 2.1.2).

3.3.1.2 Wall Effects

When charged particle tracks are passing through condensed matter that includes a gas cavity the density difference will affect energy deposition in the cavity. Four types of influence, generally known as 'wall effects', have been described by the ICRU and are schematically illustrated in Figure 3.2 from ICRU (1983). In the examples shown it is assumed that in the homogeneous situation there is no density difference between sites. It is also assumed that the atomic composition of the gas and solid matter is the same. As will be discussed in more detail in the text that follows, for the four situations illustrated, the particle can deposit energy in a gas cavity via two simultaneously occurring pathways while for the uniform density

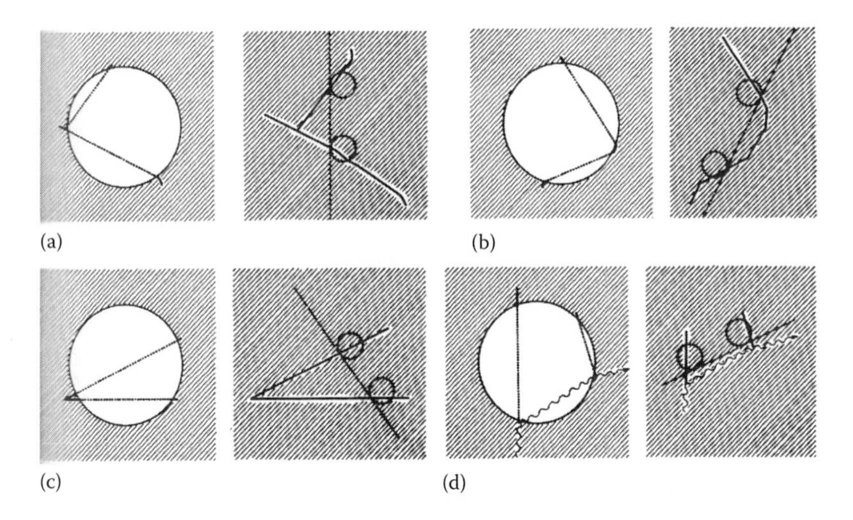

FIGURE 3.2 Different kinds of wall effects. The figures compare energy depositions by charged particles interacting with a gas cavity surrounded by a higher density material of the same atomic composition with the energy deposited in sites of the same density as the wall material. Four different situations are pictured. (a) Delta-ray effect. (b) Re-entry effect. (c) V-effect and (d) scattering effect. In all situations the gas cavity will observe an extra track as compared to the homogeneous matter. (From ICRU (International Commission on Radiation Units & Measurements). Report 36: Microdosimetry. Bethesda, MD: ICRU (1983).)

situation the two pathways would result in energy depositions in two separate sites. Thus for these wall effects it is expected that the number of events becomes smaller and the mean lineal energy larger for a gas cavity introduced in solid matter compared to the uniform density situation (Glass and Gross, 1972; ICRU, 1983; Kellerer, 1971a).

In Figure 3.2a the 'delta-ray' effect is pictured. A charged particle has created a delta electron before it enters the gas cavity. The delta electron was, however, released so close to the cavity that both particles are entering the gas volume. If there had been no density difference only one of the two particles would have deposited energy in the volume observed. These events have been referred to as *direct events* and will cause larger energy deposition events in the gas cavity than would be seen in the homogeneous density case. The 'delta-ray' effect, can be expected with high-energy heavy charged particles as well as high-energy electrons. A variant of this effect appears when the heavy charged particle passes outside of the gas cavity but sufficiently close to make it possible for delta electrons to enter the cavity. This is referred to as an *indirect event*.

The situation in Figure 3.2b illustrates what is called the ‘re-entry’ effect. In this case an electron, after crossing the gas cavity, may be scattered back into the cavity due to its curled track. With no density difference a re-entry of the site is less likely. Again, this results in a larger event in the gas cavity (walled counter) compared to the homogeneous density wall-less counter.

The effect in Figure 3.2c is called the V-effect and may arise after non-elastic nuclear reactions. If the two nuclear fragments are created outside of the cavity, both fragments may reach the cavity and simultaneously contribute to the energy imparted, while without density differences only one fragment would reach the volume of interest. The resulting effect is a large event recorded in a walled counter compared to a wall-less (homogeneous density) counter.

The fourth situation shown in Figure 3.2d applies to photons and neutrons and is called the ‘scattering’ effect. Here uncharged particles may interact through multiple scattering with atoms and may create more than one charged particle. With a gas cavity these charged particles may be recorded simultaneously, while in the uniform density case it may be more likely that only one of the charged particles will cross the site. This will also result in a larger event in a walled counter compared to a wall-less one.

The magnitude of the wall effect varies with the type of particle, its energy and the size of the simulated volume and has been studied through calculations as well as experiments. Kellerer (1971b) estimated that $\bar{y}$ would increase by about 20% in a walled counter in low-LET beams of x-rays and somewhat less at higher photon energies due to the ‘re-entry’ effect. Eickel and Booz (1976) showed through measurements with a walled and wall-less cylindrical counter that $\bar{y}$ was increased by about 40% for 200 kVp x-rays for site diameters ranging from 0.3 to 3 μm and $\bar{y}_D$ by about 10%. For a ^{60}Co γ beam the corresponding mean values were 3% and 2% in $\bar{y}$ and $\bar{y}_D$ respectively. However, Eickel and Booz caution that the differences observed are due not only to the wall effect but also to the inhomogeneity that exists between A-150 plastic and tissue equivalent gas, particularly for low-energy photons. The effect on $\bar{y}_D$ in a neutron beam of 8.5 MeV was experimentally observed to be 14% and in a 15-MeV neutron beam 8% at 1 μm (Menzel et al., 1982).

Calculations of wall effects for protons, alphas as well as ions have been reported by Nikjoo et al. (2002, 2007) and some results are shown in Table 3.3. The ions studied in the calculations were ^{20}Ne with 55 MeV u^{-1} and an LET of 132 keV μm^{-1}, ^{40}Ar with 45 MeV u^{-1} and an LET of 514 keV μm^{-1} and ^{56}Fe with 1.05 GeV u^{-1} and an LET of 138 keV μm^{-1}. In

TABLE 3.3 Wall Effects Calculated as Ratios of $\bar{y}$(Walled)/$\bar{y}$(Wall-less) and $\bar{y}_D$(Walled)/$\bar{y}_D$(Wall-less) for Protons, Alpha Particles, ^{40}Ar ions and ^{20}Ne Ions

Particle	Energy	L (keV μm^{-1})		$\bar{y}$ Ratio (Walled)/(Wall-Less)	$\bar{y}_D$ Ratio (Walled)/(Wall-Less)
		L_T	L_D		
Proton	1 MeV	26.4	26.7	1.04	1.03
	10 MeV	45	26.7	1.01	0.96
	100 MeV	0.7	0.9	1.12	0.73
	1000 MeV	0.23	0.3	0.3–0.7	4–7
α	2.4 MeV	156		1.11	1.01
^{20}Ne ion	46 MeV u^{-1}	150		3.4	1.20
^{40}Ar ion	45 MeV u^{-1}	514		28	1.34

Sources: Nikjoo H. et al., *Radiat. Res.* 157, 435–445 (2002); Nikjoo H. et al., *Radiat. Prot. Dos.* 126(1/4), 512–518 (2007).

Note: LET values (L_T and L_D) for protons are from Watt (1996). The mean lineal energies for protons of 1 GeV were given with only one significant number indicating a larger uncertainty in the calculation. For this reason, a range of values for the size of the wall effect is presented in the table. The calculations for the wall-less situation were made for a sphere of liquid water with 1 μm diameter while the walled counter had a 3 μm simulated diameter and a wall thickness of 2 μm.

both the walled and wall-less counter 80% of the energy imparted was deposited by the primary particle trajectory and its delta electrons originated inside the cavity. In the walled counter delta electrons generated by tracks in front of and behind the cavity were each responsible for almost 10% of the energy imparted and only a very small contribution came from delta electrons from ions passing outside of the cavity (indirect effect). This was contrary to the situation in the wall-less counter. For the ^{40}Ar ions 17% of the energy imparted was deposited by indirect events (delta electrons from ions passing outside the gas cavity). Delta electrons from in front and from the posterior wall contributed fractions of only single percentages. Obviously, the number of low-LET events becomes larger in the wall-less counter and $\bar{y}$ is lower. Experimental results with walled counters for a ^{56}Fe ion beam with 1.05 GeV u^{-1} by Rademacher et al. (1998) and for protons of high energies by Borak et al. (2004) are in general agreement with the calculations of Nikjoo et al. (2002). Charged particles of more than 500 MeV u^{-1} passing just at the interface of the wall and cavity will create large events in the cavity due to the increased number of delta electrons created in the wall (Kellerer, 1971b; Nikjoo et al., 2002; Rademacher et al., 1998). Below 500 MeV u^{-1} the importance of these wall-generated delta electrons is less. These very large events seem not to affect $\bar{y}_D$ much, but will be visible in a single-event distribution as very large events above the expected event distribution.

From the preceding it is clear that if precise values of the frequency mean and dose mean lineal energies are required it becomes important to measure with wall-less counters. Two types of wall-less detectors have been in use. The most successful has been the grid detector, which is essentially a Rossi counter with a grid of A-150 TE plastic replacing the wall. The transparency of the wall is typically 90%. In the second technique, specially designed electrodes create electric field lines that define a spherical collecting volume in the space between them (Glass and Gross, 1972).

3.3.1.3 Gas Gain

The electric current as a function of the applied voltage in a gas ionisation device for a constant radiation fluence is shown in Figure 3.3. At very low voltages, the current increases rapidly to an almost constant level, after

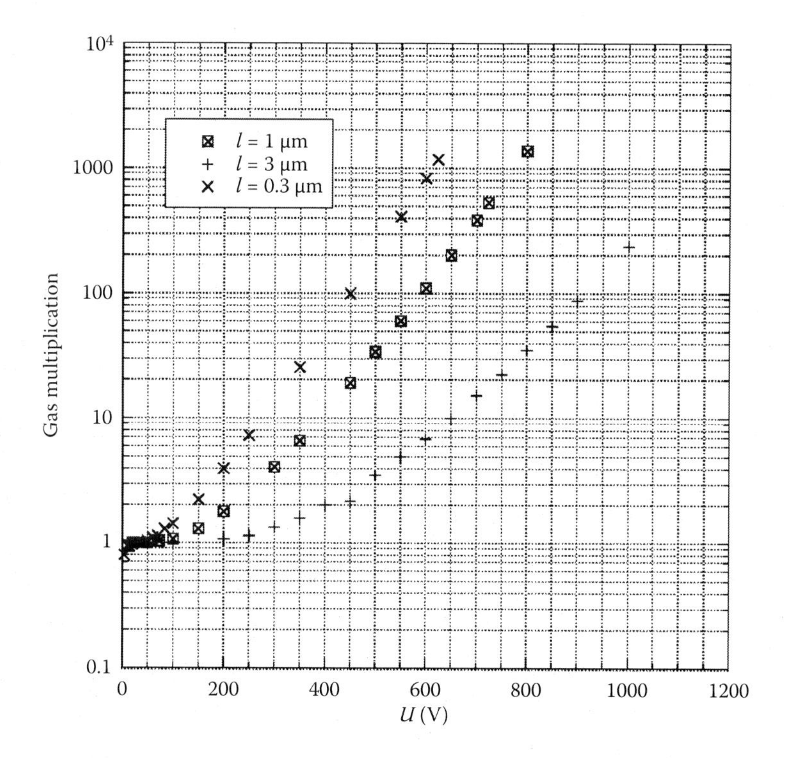

FIGURE 3.3 Gas multiplication measured in a cylindrical TEPC at different potentials and three simulated site volumes. (The detector design is described in Lindborg et al., 1985.)

which it again begins to increase. In the initial phase the electric field strength is so small that recombination between ions and electrons is highly probable. In the first plateau region, the electric field has become strong enough to make the collection of the created ions almost complete. This plateau defines the ionisation chamber region. The remaining recombination can be determined experimentally. There are two different types of recombination: initial and volume recombination. The first occurs between electrons and ions within the same track, while the latter occurs between ions and electrons from different tracks. In single-event measurements the initial recombination is the only effect, while in multiple event measurements both phenomena may be present. It is unusual that this effect needs to be corrected for except in the case of variance measurements at high dose rates where a correction of a few per cent has been reported. For continuous non-pulsed radiation fields by plotting the reciprocal of the measured collected charge against the reciprocal of the applied potential squared and extrapolating to zero reciprocal charge, that is, infinite applied voltage, a correction for volume recombination may be made (Boag, 1966).

Initial recombination depends on the proximity of the ions formed in particle tracks and therefore depends on radiation quality. This phenomenon has been used by a number of workers as the basis of experimental microdosimetry using high-pressure ionisation chambers to determine quality factors or quantify restricted LET, as will be discussed later in this chapter.

Returning to the ionisation current characteristic of Figure 3.3, when the applied potential increases beyond the saturation region the current again increases due to gas multiplication. In this range, there is proportionality between the initial number of ion pairs created along the track and the measured number of electrons at the chamber anode. This voltage range defines the proportional counter region.

Many descriptions of the gas multiplication process are available in the literature, such as Knoll (2010). Each scheme has its merits for describing the performance of proportional counters, at least in a specified range of pressures and reduced electric field strengths. As will be understood following the discussion on TEPC calibration with the PHA method, we do not need an absolute value for the gas gain of a TEPC to carry out experimental microdosimetry; however, some appreciation of the multiplication process and how it is affected by the design of the counter including the gas fill and applied potentials is useful for an overall understanding of TEPC design and performance. For this purpose, we can apply the classical analysis of gas gain originally proposed by Townsend and applied to low-pressure proportional counters by Campion (1971).

The classic description of gas multiplication involves the Townsend first ionisation coefficient, α, which is the number of ionisations a particle causes when travelling a unit path length. If N_0 primary particles pass an infinitesimal small distance dx the number of secondary electrons created is

$$dN = N_0 \alpha dx \tag{3.14}$$

Integration over a distance d gives the number of electrons, N, created after passing this distance and

$$N = N_0 e^{\alpha d} \tag{3.15}$$

The gain G is then defined as

$$G = \frac{N}{N_0} = e^{\alpha d} \tag{3.16}$$

The coefficient α is assumed to depend on the pressure P and the electric field strength E according to

$$\frac{\alpha}{P} = Ae^{-BP/E} \tag{3.17}$$

The constant A is proportional to the reciprocal of the mean free path in a gas at a given pressure, while B is related to the ionisation potential of the molecules in the gas (von Engel, 1994). A and B are gas specific and for methane-based gas approximate values of the two constants are A = 10 cm^{-1} torr^{-1} and B = 210 V cm^{-1} torr^{-1} (Campion, 1971). Waker (1982) reported values for A and B for propane-based TE gas of 34 cm^{-1} torr^{-1} and 418 V cm^{-1} torr^{-1} determined with TEPCs simulating site sizes ranging from 1 to 10 μm.

If the logarithm of both sides in the preceding relation is derived, the relation becomes

$$\ln\frac{\alpha}{P} = \ln A - \frac{B}{E/P} \tag{3.18}$$

The reduced Townsend coefficient, α/P, is expected to become constant when E/P gets large and $\alpha/P = A$. This is not quite in agreement with observations and a better agreement has been found at low E/P when α is directly proportional to E, while at $E/P > 7.5$ V Pa^{-1} cm^{-1} a proportionality between α and $E^{1/2}$ is observed (ICRU, 1983). Ségur et al. (1995) have theoretically shown that at reduced gas pressures the electrons may start to orbit the anode without being collected by it and this may be one reason for the decrease in α, which means that the gas multiplication becomes less efficient. The uncertain physics at very high reduced electric field means that extreme care must be taken in evaluating single-event measurements determined for sub-micron simulated site sizes.

Usually the electric field strength varies in a counter, and Equation 3.16 is replaced by

$$G = e^{\int_c^a \alpha(x)dx} \quad (3.19)$$

Here c is the point of ionisation and a is the point of collection (the anode for electrons).

The electric field strength in a cylindrical counter is given by

$$X = \frac{V}{\ln \frac{c}{a}} \quad (3.20)$$

Here c and a are the radii of the cathode and anode and V is the applied potential. The gain when an electron is transported between the two positions is then

$$\ln G = \int_a^c \alpha\, dr = \frac{AV}{B \ln \frac{c}{a}} \left[e^{-aP\frac{B\ln\frac{c}{a}}{V}} - e^{-cP\frac{B\ln\frac{c}{a}}{V}} \right] \quad (3.21)$$

As $c >> a$ the second term within the brackets may be ignored and the relation can be approximated by

$$\ln G \approx \frac{AV}{B \ln \frac{c}{a}} \left[e^{-aP\frac{B\ln\frac{c}{a}}{V}} \right] \quad (3.22)$$

or after expansion as a series

$$\ln G \approx \frac{AV}{B \ln \frac{c}{a}} \left[\frac{c}{a} \right]^{\frac{-aBP}{V}} \quad (3.23)$$

This also provides a very simple test for a TEPC with a built-in calibrating source as the gas gain can be determined by observing the magnitude of the α peak as a function of applied voltage. A plot of the logarithm of the gas gain against voltage should be to a good approximation a straight line, and significant deviation from linearity is an indication that the proportional counter region may not have been reached or may have been exceeded. This linearity is observed in Figure 3.3 for the higher applied voltages.

3.3.2 Ionisation Chambers

Ionisation chambers have been used in the VC method. A great advantage with these detectors is that they are commercially available in many forms and sizes. They are already prepared for electrometer measurements and in general furnished with an electric guard, which reduces leakage currents, which is a great advantage in low-current measurements. Wall effects will appear in the same way as in proportional counters. Many commercial ionisation chambers are equipped with a large central charge collecting electrode, which will increase the secondary electron emission (SEE) current (Section 3.4.2.2), and may have an influence on the wall effect as well. TE plastic is less frequent as a wall material in ionisation chambers. Grindborg et al. (1995) reported measurements in photon beams using ionisation chambers made by the air-equivalent material C-552 and filled with air. The measured dose-mean energy imparted in air was subsequently converted to a value for tissue.

3.3.3 Recombination Chambers

It has already been mentioned that the recombination of ions in an ionisation chamber depends on the proximity of the formation of the ions and therefore will be related to radiation quality. Recombination chambers are just high-pressure ionisation chambers where the ionisation current is determined with high accuracy for two different collecting voltages and the difference in measured current related to the quality of the radiation. The concept of using ion-recombination in microdosimetry was first introduced by Sullivan (1968) and Dennis (1968) in the First Symposium on Microdosimetry in 1967 and later developed by Makrigiorgos and Waker (1986). The technique had earlier been used by Zielczynski (1962) for relative biological effectiveness (RBE) determinations. In the development of recombination chambers for microdosimetry there have been two main approaches. One has been the use of empirical relationships between measured recombination losses and quality factors used in radiation protection (Golnik, 1996) and the other in which the degree of recombination is linked through a theoretical description of the process to the dose mean restricted LET (Makrigiorgos and Waker, 1986).

The advantages of the recombination method are similar to those of the variance technique and measurements can be made in radiation fields where the dose rate is too high for pulse height measurements. Furthermore, initial recombination takes place over dimensions equivalent to tissue volumes of about 75 nm in size. In a number of papers Golnik and co-workers have demonstrated over the past couple of decades the ability of recombination chambers to be practical field instruments for use in mixed-field radiation protection dosimetry (Gryzinski et al., 2010). Earlier theoretical developments in recombination theory were carried out without the use of Monte Carlo track-structure codes and this opens up a research opportunity to put recombination chambers on a firmer theoretical foundation for use in experimental microdosimetry if the appropriate event–by–event Monte-Carlo code for the gases used in recombination chambers were to be developed.

3.3.4 Gas Electron Multipliers

Gas electron multipliers (GEMs) are a relatively new form of charged particle detector utilising internal signal amplification through a gas-avalanche mechanism. This device was first described by Sauli in 1977 with further developments reported in Sauli (2003, 2010). A typical GEM consists of two 5 μm thick layers of copper separated by a 50 μm thick insulating Kapton® foil. The GEM foil is chemically pierced with a high-density pattern of 50 to 100 holes per mm^2. Individual holes are bi-conical in shape with entrance and exit diameters of 70 μm, waist diameter of 50 μm and hole pitch 140 μm; see Figure 3.4a.

The principle of operation of the GEM is that a drift field is established above the GEM foil by using the upper surface of the GEM as the anode and an external electrode as the cathode. Any ionisation in the gas between these two electrodes will create ion pairs with the electrons drifting towards the upper GEM surface. The drift region should operate as an ion chamber with no recombination of the created ion pairs. The upper and lower copper surfaces of the GEM foil are held at different electrical potentials and so the free electrons created in the drift region are guided into the GEM holes, where they experience a strong electric field created by the potential difference between each side of the foil and the thinness of

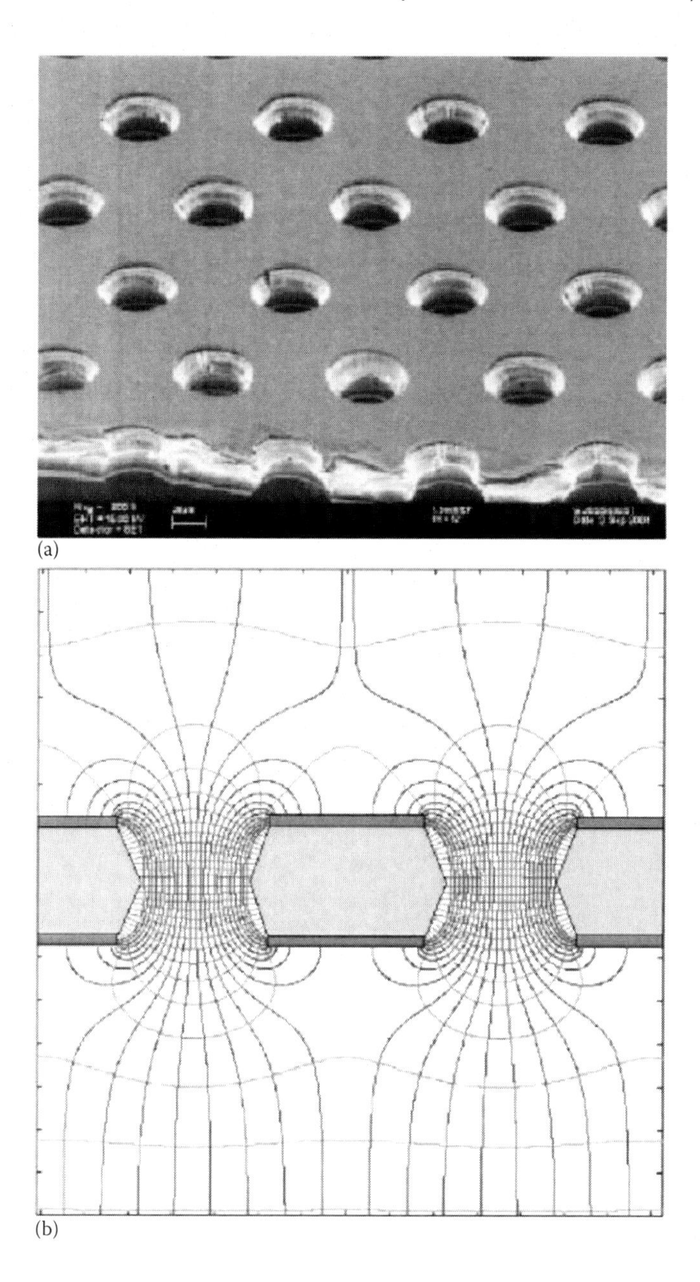

FIGURE 3.4 (a) The general construction and layout of a GEM foil (b) as well as details of the electric field distribution through the etched holes which enables gas multiplication. (Images courtesy of The Gas Detectors Development Group, European Organisation for Nuclear Research (CERN).)

the foil itself (Figure 3.4b). With the correct potential across the GEM foil the electric field within the hole is sufficient that electrons entering the hole can undergo multiplication. Thus each hole acts as a miniature proportional counter. As the geometry of each hole is identical and the voltage across the GEM is uniform the electric field and gas gain within each hole will be the same. On exit from the GEM foil the electrons are collected by a further electrode, the collector plate, situated beneath the GEM. A potential difference is also established between the lower GEM surface and the collector electrode in order that electrons generated within the GEM continue to drift towards and be collected by the collector plate. In principle, GEM devices can be used for microdosimetry provided the gas drift region above the GEM simulates a unit density tissue volumes of microscopic dimensions. The main perceived advantages of GEMs for microdosimetry are their compactness and the flexibility in design of the collector electrode. Compactness is important concerning the possibility of designing TEPC type instruments for personal neutron and mixed field dosimetry or for operation in high-intensity fields. Flexibility in collector electrode design could lead to the ability of tracking individual charge particles and therefore particle discrimination. Farahmand (2004) demonstrated that a counter volume 1.8 mm diameter filled with propane TE gas simulating a 1-μm tissue volume and using a GEM for gas multiplication was able to successfully measure single-event lineal energy distributions for 14 MeV and Cf-252 neutron fields. Dubeau and Waker (2008) showed that by using a lithographically produced microstrip readout pattern for the collector electrode (see Figure 3.5), they were able to obtain the equivalent of a large number of miniature TEPC detector elements and achieve a sensitivity in terms of counts per micro Sievert that indicated potential for a personal neutron monitoring device. One of the difficulties of GEM detector development is the limited range of size and shape of the foils and the very limited number of suppliers available. A recent innovation in GEM technology that somewhat offsets these difficulties is the so-called thick GEM (THGEM) first reported by Periale et al. (2002) and later further developed by the detector group at the Weizmann Institute of Science (Chechik 2004, 2005). The principle of operation of the THGEM is the same as the GEM described earlier; however, the diameter of the holes and thickness of the insulator are much greater, making the device more robust and easier to fabricate. Typically, in a THGEM the insulator thickness is of the order of a 100 μm or more and hole diameters of around 300 μm. Orchard et al. (2011) have reported

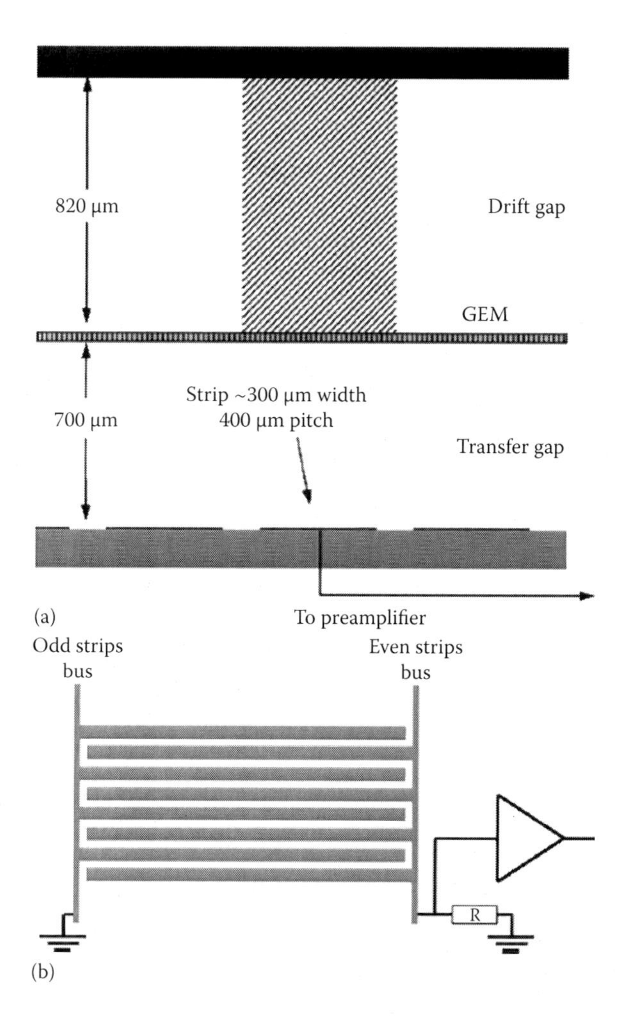

FIGURE 3.5 A schematic of the GEM-based TEPC designed by Dubeau and Waker (2008) (a) and the collecting electrode pattern (b) that enabled the device to be operated as an ensemble of miniature TEPCs.

the operation of a THGEM as a TEPC in which the insulator was 120 μm thick, coated on each side by copper 60 μm thick and with 32 holes each with a diameter of 350 μm and a pitch of 640 μm. A narrow rim was created around each hole by etching back the copper coat about 0.1 mm. The drift region above the THGEM was a 5 mm × 5 mm right cylinder representing a 2 μm tissue volume when filled with propane TE gas at 167 torr (22.26 kPa). With this device Orchard et al. (2011) reported linearity of the logarithm of the relative gas gain and sufficient gas gain to measure event sizes down to around 1.5 keV μm^{-1}, sufficient to measure both neutron and photon components in a mixed neutron-gamma field. Further investigation and characterisation of THGEMs as TE proportional counters is required, but their future contribution to experimental microdosimetry, as is of the case for regular GEMs, looks promising.

3.3.5 Diodes

The potential advantages of using solid-state detectors such as p–n diodes for microdosimetry are numerous and include the fact that they can directly provide sensitive volumes of micrometre dimensions and thereby are much more suitable for measurements in intense beams such as used in radiation therapy. Diode detectors can be manufactured fairly cheaply, operate at low voltage and power consumption and are not sensitive to vibrations and microphonics. Having a solid-state detector of microscopic dimensions also avoids wall effects encountered by simulating unit density tissue by a low-density gas volume contained within a solid-walled shell. These advantages were recognised more than three decades ago; however, it is only recently that the full potential of solid-state microdosimetry has begun to be realised. A review of the concepts and applications of silicon detector based microdosimetry has been given by Rosenfeld (2016), which should be consulted for a comprehensive introduction to this novel detector technology.

In using silicon devices for microdosimetry there are two principal difficulties encountered, one being the exact delineation of the sensitive volume and the other the lack of tissue equivalence of silicon; both of these issues are being solved in innovative ways. The problem concerning the exact boundaries of the sensitive volume arises from the so-called 'field-funnelling effect' in which the passage of a high-LET particle across the depletion layer of the diode can distort the local electric field sufficiently so that charge can be collected from outside the depletion region, thereby effectively increasing the site size of the microdosimeter. To overcome this problem silicon on insulator (SIO) devices (Prokopovich et al., 2008) have been developed enabling charge collection regions and therefore sensitive volumes of well-defined dimensions to be fabricated.

Tissue equivalence in silicon devices presents itself as an issue in two ways, one concerns the interaction of radiation with the detector itself and the other with the deposition of energy by charged particles in a material (silicon) that has a different mass stopping power than tissue. The first problem for neutron irradiation is readily solved by using a TE or hydrogen-rich material (polyethylene, for example) in contact with the silicon device surface to act as a converter so that the diode detects the elastically scattered protons from the primary interaction of neutrons in the converter. Non-tissue equivalence of the silicon diode forming the microdosimetric sensitive volume has to be corrected by applying a stopping power ratio of tissue to silicon for the charged particles imparting energy to the diode. A development that enables this correction to be made is the use of monolithic silicon telescopes (Agosteo and Pola, 2011). In the telescope design there is a micrometre ΔE region which acts as the microdosimeter and an E region that measures the residual energy of the charged particle crossing the sensitive volume defined by the ΔE layer. The two stages of charge collection are fabricated on a single silicon substrate separated by a deeply implanted p-region, which separates the two volumes and their charge collection regions. Energy deposition in the ΔE region together with the E region energy deposition enables the ion responsible for the event to be identified and the correct stopping power correction between silicon and tissue to be made. For very energetic protons Agosteo and Pola use an average value of 0.574 for this stopping power ratio. Lower energy protons that deposit all their remaining energy in the E section of the telescope enable an event-by-event correction to be made by identifying the energy of each proton and using the appropriate value for the stopping power ratio between tissue and silicon.

Agosteo and others further improved performance by creating an array of 'telescopes' so as to maintain sensitivity, but also avoiding other geometric effects they found related to sensitive volume definition and the relatively wide surface area of the original detector design (Agosteo et al., 2011). The array they reported consisted of 7000 pixels connected in parallel to give an effective collection area of about 0.5 mm^2. Each pixel consisted of a ΔE region 2 μm thick by 9 μm in diameter and an E region 500 μm thick. Figure 3.6 shows the agreement between the 'array' detector when compared with a cylindrical TEPC for various neutron energies. Figure 3.6 also shows a further problem with diode detectors used for microdosimetry and that is their inability to measure event sizes below values around a few 10 keV μm^{-1} due to the electronic noise generated by the high capacitance of the micrometre thick sensitive zone; this is a significant limitation for photon, mixed-field and high-energy proton microdosimetry.

In spite of the various shortcomings of silicon diodes for microdosimetry there is little doubt, however, that the advantages of these

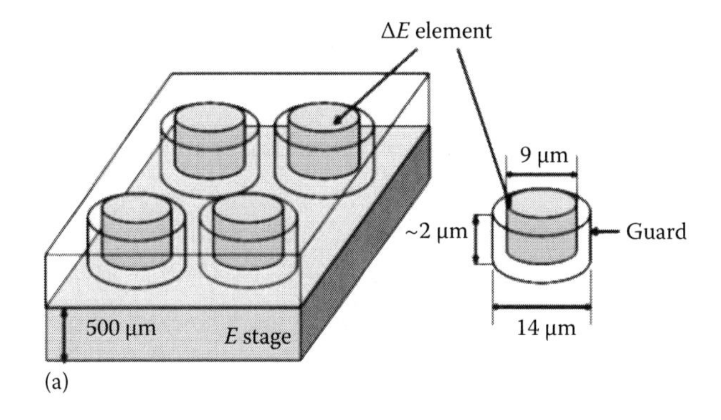

FIGURE 3.6 A schematic of the silicon telescope microdosimeter developed by Agosteo and Pola (2011) (a). *(Continued)*

detectors will ensure their future place as an important innovation in experimental microdosimetry, especially in high-LET radiotherapy beam characterisation or in space radiation protection dosimetry.

3.3.6 Diamond Detectors

Another potential solid-state detector for experimental microdosimetry that has emerged is the chemical vapor deposition (CVD) diamond detector (Rollet et al., 2012). This detector consists of a high-purity intrinsic monocrystalline diamond detecting region of thickness less than 5 µm, the sensitive volume, grown over a backing of boron doped monocrystalline diamond that acts as a deep conductive electrical contact. To date this type of detector has been experimentally tested with alpha particles and the results of measurements compared with Monte Carlo simulations in order to examine the potential of this detector type for quantifying radiation quality of therapeutic ion beams (Solevi et al., 2015).

3.4 MEASUREMENT METHODS

3.4.1 Pulse Height Analysis of Single Events

3.4.1.1 Measurement Principle

The most common experimental method for microdosimetry is based on pulse height analysis (PHA) as illustrated in Figure 3.7. The technique is very much the same as used in gamma-ray spectrometry measurements to identify nuclides in a sample.

The major differences are the many small energy deposition events (pulses) that have to be recorded, and the extensive range of pulse size

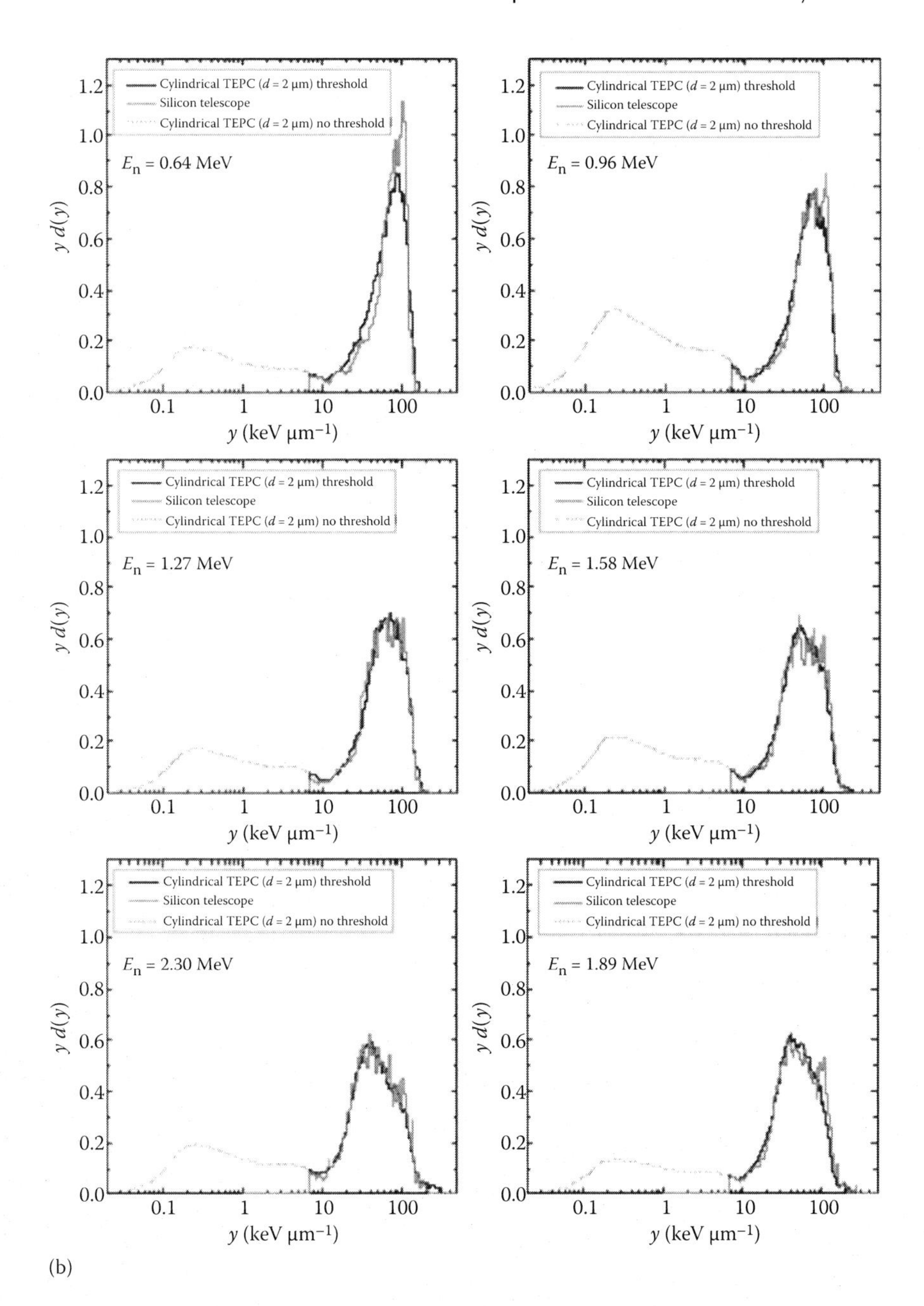

FIGURE 3.6 (CONTINUED) Results of measurements carried out at various neutron energies with the Si telescope and a conventional cylindrical TEPC (b).

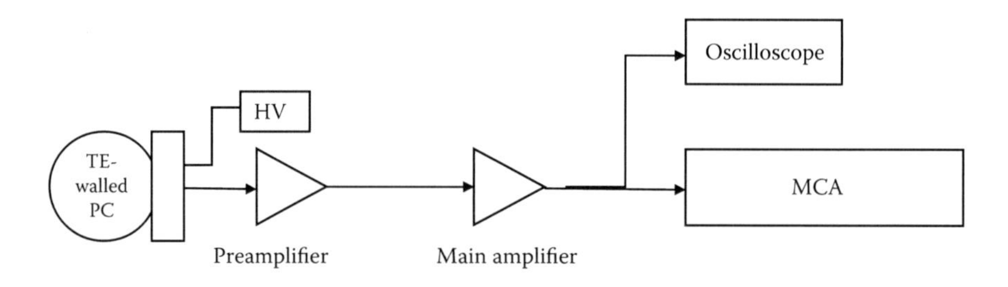

FIGURE 3.7 Block diagram of a simple set-up for PHA. Each energy deposition event in the TE proportional counter is converted to a voltage pulse by the charge sensitive preamplifier, shaped and amplified by the main amplifier and the resulting pulse height recorded by the MCA.

between the largest and smallest event size generated in a TEPC. To master the latter, measurements are usually performed with different amplifier gain settings either by having, for instance, three different linear amplifiers followed by three different analogue-digital-converter (ADC) connected to the multi-channel analyser (MCA), or sequential measurements repeated three times with three different settings of the amplifier gain. This latter method minimises the equipment needed but requires longer measurement times. The three different spectra must then be combined in post-measurement data analysis; this can be done either manually or by computer. If instead of linear amplifiers a logarithmic amplifier is available, the whole pulse height spectrum can be obtained in a single measurement. In both cases events below 0.1 keV μm^{-1} are usually not possible to resolve accurately and spectra have to be completed by extrapolation down to say 0.01 keV μm^{-1}. Without extrapolation the value of $\bar{y}$ is affected by as much as 30% in a high-energy photon or electron beam. The effect on $\bar{y}_D$ is typically 5% in a low-LET beam and less in a high-LET beam. In commercially available instruments, the extrapolation may be done automatically and the methodology should be understood to avoid unnecessary errors in calculating microdosimetric quantities (Lillhök et al., 2007c). A comparison of dose distributions measured with commercial TEPC-based instruments and Monte Carlo–calculated (FLUKA) dose distributions have been reported by Beck et al. (2004) and illustrates the issue of lower level measurement thresholds and extrapolation. Figure 3.8 further shows that extrapolation from a lower level threshold of around 0.6 keV μm^{-1} will still underestimate the photon dose and overestimate $\bar{y}$. If the measurement of the photon

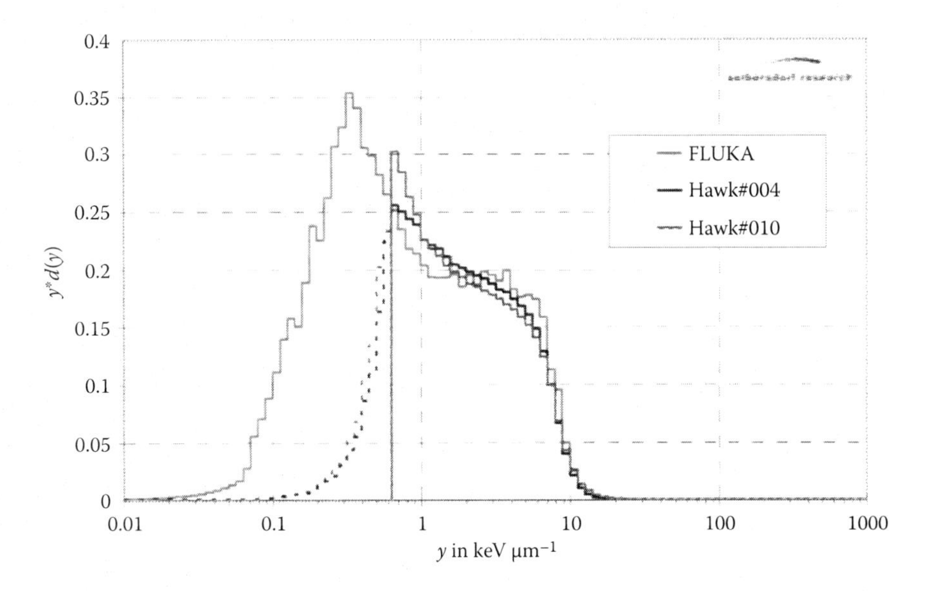

FIGURE 3.8 Single-event dose distributions measured in a ^{60}Co γ field with two spherical commercial TEPCs (Hawk instrument, Far West Technology). A FLUKA MC-code calculated dose distribution for a TEPC is shown for comparison. Both Hawk instruments had a lower threshold at about 0.6 keV μm^{-1}, while the calculation included energy depositions below this level. The experimental distributions were extrapolated linearly to 0.01 keV μm^{-1}, dashed lines. (From Beck P., Latocha M., Rollet S., Ferrari A., Pelliccioni M. and Villari R. In L. Lindborg, D. T. Bartlett, P. Beck, I. R. McAulay, K. Schnuer, H. Schraube and F. Spurny (eds.), *Cosmic radiation exposure of aircraft crew: Compilation of measured and calculated data*, Appendix A. Radiation Protection Issue No. 140. European Commission (2004).)

dose or the photon dose component of a mixed field is important then every effort should be made to measure lineal energies down to 0.1 to 0.2 keV μm^{-1}.

3.4.1.2 Electronic Noise and Gas Gain Statistics

To detect the many small events a low noise preamplifier is needed. Usually a charge sensitive preamplifier with a field-effect transistor (FET) at the input is used. The noise level is also dependent on the capacitance of both detector and signal cable before the preamplifier stage.

A detector typically has a capacitance of one to a few picofarads (pF). The capacitance of a high-quality signal cable is typically 50 pF m^{-1}. As the dominant electronic noise source will increase proportionally to the input capacitance, it is clear that a short cable between the detector and preamplifier is advantageous. For this reason, the first stage of the preamplifier can be mounted directly on the detector. However, when a preamplifier is irradiated electric currents may be created and affect the measurements at least at higher dose rates. As a rule of thumb, a noise level of 130 root mean square (rms) electrons has been the lower limit achieved in practice. An rms-reading voltmeter may be useful to check the noise level. An alternative to the rms-voltmeter is to observe the broadening of the pulse from a pulse generator in the MCA. The broadening in terms of full width at half maximum (FWHM) is usually determined and used as a measure of the noise. The relation between the two measures is FWHM = 2.36 (rms).

The gas gain itself will contribute to the broadening of the measured distribution of energy imparted. The so-called single-electron spectrum (the probability that a particular number of electrons will arrive at the anode as the result of one ion pair being generated in the gas) determines the degree of broadening. The overall gain distribution results from the many-fold convolution of the single-electron spectrum depending on the number of free electrons released due to the event; therefore the uncertainty in the exact size of the event is greater the smaller the event and the number of electrons released.

From the preceding it is obvious that the lower part of a measured single-event spectrum becomes uncertain. Quite a large fraction of the events may actually be found in this region; this is particularly the case for photon beam measurements.

3.4.1.3 Representation of Measured Results

Experimental microdosimetry using the pulse height method consists of determining the frequency of events occurring with a particular lineal energy. Experimentally determined event spectra are without distinguishing features appearing largely as a decreasing exponential function. However, converting the measured frequency distribution to a so-called $yd(y)$ dose distribution introduces a number of recognisable features that provide a useful degree of visual diagnostic information. This transformation of the data is obtained by first multiplying each frequency value by the size of the event itself and then redistributing the data into equal logarithmic lineal energy intervals as was briefly discussed in Chapter 2, Section 2.5.1. Of course, the frequency information alone could be redistributed logarithmically for presentation purposes if the analysis of the experimental results required a specific discussion on the number of events rather than the energy deposited by them. However, it is usually the total energy deposited by the events rather than their number that is of interest in radiation effects and radiation protection and so it is the dose distribution that is most often displayed. The visual stages of transforming measured single-event spectra are indicated in Figure 3.9a, b and c. The exact method and software for logarithmically redistributing pulse height frequency data recorded with a linear device such as an MCA depends very much on the particulars of the MCA or histogramming memory used in the data acquisition system and therefore is something that each laboratory or research group usually has to develop themselves. A typical logarithmic redistribution process might consist of constructing a scale in which each decade of linear energy is divided into 50 equal logarithmic intervals

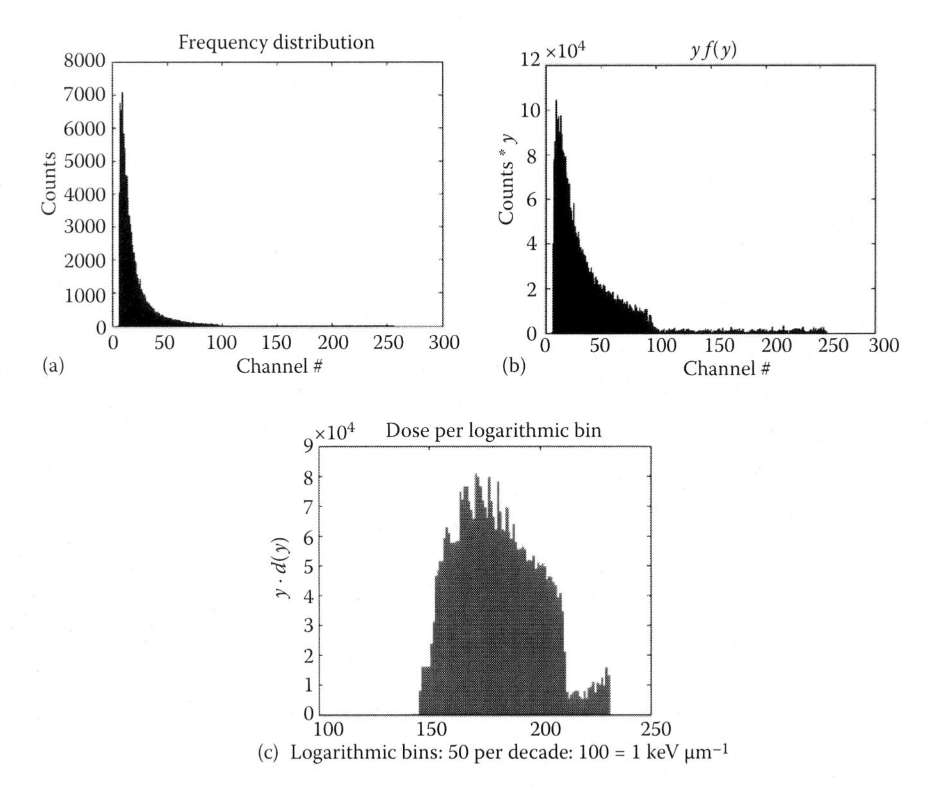

FIGURE 3.9 **(See colour insert.)** Data from a TEPC (REM500) instrument made in an Am–Be neutron field. The pulse height spectrum representing the frequency distribution (a) is converted into a dose distribution (b) by multiplying the frequency of an event by the size of the event. Finally, the dose distribution is redistributed on a logarithmic scale consisting of equal logarithmic intervals (c), in this case 50 logarithmic intervals per decade of lineal energy.

or bins. Thus an event-size spectrum covering a lineal energy range of 0.01 keV μm^{-1} to 1000 keV μm^{-1} would consist of 250 logarithmic bins. The actual logarithmic distribution process would then consist of using the lineal energy limits of each logarithmic interval to locate the same range of lineal energy in the 'frequency multiplied by the event-size' linear data set, the size of which is determined by the data acquisition system used, and summing the total number of 'counts times event size' recorded and placing them in the logarithmic bin being filled. It is important to keep in mind that the raw data of the measured frequency distribution remains the same – it is simply a matter of how these data are displayed. As stated in Chapter 2, one advantage of this method of data representation is that equal areas under the curve between any two lineal energy values represent equal contributions to the total absorbed dose. Just as useful, the structure introduced into the shape of the spectra by logarithmic re-binning of the dose distribution also enables some identification of the types and energy of particles generating the events.

As an example of this capability, Figure 3.10 shows the single-event dose distributions for several neutron energies plotted on an equal logarithmic interval scale of lineal energy. We can identify the peak in the distributions, broadly speaking between 10 keV μm^{-1} and 100 keV μm^{-1} with events from elastically scattered protons and the peak(s) due to alpha particles

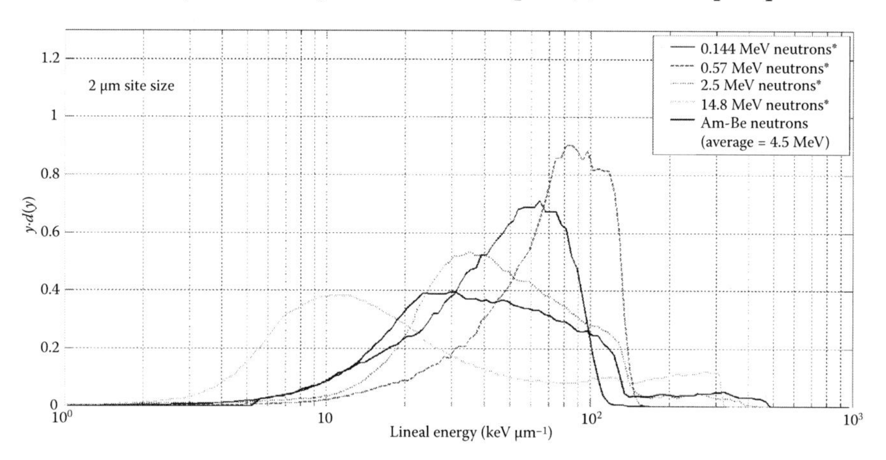

FIGURE 3.10 **(See colour insert.)** Single-event lineal energy dose distributions for several different neutron fields measured with a 5-inch spherical TEPC and displayed in the *yd*(*y*) format on an equal logarithmic interval scale. The position of the recoil proton peak and the proton edge all convey information to the experimenter that can be useful in circumstances where the exact nature of the radiation is not known a priori.

and elastically scattered ions of O, N, C in the region above 100 keV μm^{-1}. The sharp drop in the plot above around 100 keV μm^{-1}, known as the 'proton edge', results from the maximum energy that can be imparted by a charged particle (in this case a proton) to the site being simulated. This maximum in imparted energy always results when the stopping power of the charge particle is at a maximum and the particle's path across the site is also at a maximum. These conditions occur when the range of the charge particle just matches the simulated site diameter. Changes in the proton peak position as well as the number of events beyond the proton edge are all features of the spectra that can convey information to the experimenter about the nature of the radiation field, which would not have been easily identifiable if the data had not been presented in a logarithmic display. Examining Figure 3.10, a recoil proton peak seen in the region of 10 keV μm^{-1} or lower accompanied by a significant number of events in the lineal energy region above 100 keV μm^{-1} would indicate a radiation field of predominantly high-energy neutrons. However, a proton peak close to the proton edge around 140 keV μm^{-1} and very few events at higher lineal energies would signify a mainly low-energy neutron field. The position of the proton edge also provides information concerning the average energy of the incident neutrons. For a volume simulating a diameter of 2 μm the proton edge is expected to be around 139 keV μm^{-1}; if a measured proton edge is significantly lower than this value for a 2 μm measurement this is an indication that the radiation field consists of neutrons of energy less than a few hundred kiloelectron volts.

A final word regarding the logarithmic representation of microdosimetric spectra concerns 'normalisation'. Figure 3.10 is an example of the use of normalised spectra in which we can directly compare several spectra regardless of the actual quantity of data collected. With normalised spectra the total area is made equal to 1 and equal areas under the curve now represent equal dose fractions of the total dose. That is,

$$\text{Area} = \int_{\min}^{\max} yd(y)d\ln y = 1$$

To achieve this normalisation the individual values of $yd(y)$ plotted must be corrected by dividing by the sum of all $yd(y)$ values and the magnitude of the logarithmic interval. The numerical value of the logarithmic interval can be found from the number of logarithmic intervals used per decade of lineal energy. Thus if 50 intervals are used then $d\log y = \frac{1}{50}$ and $d\ln y = \frac{\ln 10}{50} = 0.04605$; generalising, if B logarithmic bins are used per decade of lineal energy then $d\ln y = \frac{\ln 10}{B}$.

3.4.1.4 Calibration Methods for Lineal Energy

There are a few different calibration methods for TEPCs and the PHA method reported in the literature. They are all based on the mean energy imparted by a particle of known LET crossing the gas cavity or particles having a known energy that is entirely deposited in the TEPC gas cavity. This deposited energy is then associated with a channel number on a MCA for a particular amplifier gain setting and applied anode voltage.

3.4.1.4.1 Alpha Source Calibration The most common method is to use an alpha source with a well-known energy that is mounted outside the wall of the TEPC. In its normal position the source is blocked by a shutter and alpha particles cannot pass through the wall of the TEPC and cross the gas cavity. However, either by using a magnet to lift the shutter or by rotating the counter and using gravity to drop the shutter, alpha particles can be collimated through an opening in the TE wall and directed across the diameter of the counter. With a known value for the energy of the alpha particles entering the cavity and for a known simulated diameter the energy lost by the alpha particle in crossing the counter can be calculated using range-energy data for the gas being used. For example, for a Cm-244 alpha source with an alpha particle energy of 5.76 MeV the energy deposited in crossing a 2 μm unit-density propane-based TE gas volume, ΔE, is 170.05 keV, which gives a calibration factor of 127.5 keV μm^{-1} in lineal energy with the mean chord length taken as two-thirds the diameter in a spherical counter.

$$y_\alpha = \frac{\Delta E}{\bar{\ell}} = \frac{170.05}{2} \times \frac{3}{2} = 127.5 \tag{3.24}$$

The relation between the channel number on the MCA where the alpha pulse height peak appears and y_α yields a calibration factor in terms of keV μm^{-1} per channel. Waker (1985) has pointed out that one significant component of uncertainty in using the alpha source method for calibration is the alignment of the source and therefore not knowing the exact trajectory of the alpha particles across the counter. This was investigated by comparing values of $\bar{y}$ and $\bar{y}_D$ for three TEPCs calibrated using alpha particles and the so-called 'proton edge' method described in the text that follows. Proton edge calibration leads to an improvement in the coefficient of variation of the three measured $\bar{y}_D$ values from 8.5% to 0.6%. As the alpha sources are often coated with an ultra-thin layer of material such as gold in order to stabilise them, knowing their exact initial energy can also be an issue. In constructing a new counter in which an alpha source is to be used for calibration purposes it is worthwhile measuring the mean energy of the emitted alphas by alpha spectrometry prior to use.

Internal alpha sources are not only a convenient method of calibration; they also enable an easy method of checking gas gain stability and resolution of a TEPC and it is unfortunate that commercially obtained TEPCs are now usually sold without an internal alpha calibration source. Even for detectors built in house it is becoming difficult to purchase small Am-241 sources to use for TEPC calibration; consequently it is increasingly necessary to turn to intrinsic methods of counter calibration such as using the 'proton edge' or 'electron-edge'.

3.4.1.4.2 Proton Edge Calibration The energy that a charged particle can deposit in a microscopic volume results from the product of the particle's path length and LET, that is $\Delta E = \frac{dE}{dx}\Delta x$, and the maximum deposited energy possible will be when both Δx and $\frac{dE}{dx}$ are at their maximum values, which means the particle crosses the diameter for a spherical counter with an energy such that its range matches the simulated site size with the maximum stopping power or LET being at the end of the particle's path. Using a charged particle 'edge' to calibrate a TEPC consists of first identifying the edge and then ascribing the correct lineal energy to its position in the event-size spectrum. The proton edge is a very identifiable feature for event-size spectra measured in neutron fields, but the difficulty in using this method of calibration is knowing what lineal energy corresponds to the 'edge' for the simulated diameter being used. Some workers take the middle point of the drop in event size and others extrapolate the edge and find the position at which it meets the lineal energy axis. Waker, for example, has suggested extrapolating the edge to the lineal energy axis and denoting that position to be 139.5 keV μm^{-1} for propane TE gas and a 2-μm simulated site size; however, this will depend on what proton stopping power values are used and how these values are integrated as the proton crosses the counter and changes energy (Waker, 1985). Uncertainty in calibration may well be of the order of a few per cent, which is not likely to be an issue in radiation protection measurements but certainly could be for the higher precision work required in therapy beam dosimetry.

3.4.1.4.3 Electron Edge Calibration Similar to any other charged particle, electrons will also create an 'edge' in a single event spectrum which will be around a lineal energy value of 10 keV μm^{-1} for site sizes of 1 to 2 μm and for photon fields such as ^{137}Cs and ^{60}Co. Using the electron edge as means of calibrating TEPCs was investigated and formalised by Crossman and Watt (1994) and has recently been revisited and further investigated by Moro et al. (2015). Moro and co-workers described the most precise way of identifying the position of the electron edge and then, through a series of experiments with TEPCs, operated with different simulated diameters, were able to systemise the relationship between the most appropriate lineal energy value for the edge and the site size simulated by the TEPC. Direct evaluation of the lineal energy corresponding to an electron edge has a large uncertainty due to uncertainties in electron stopping power data at the energy of the electrons that contribute to maximum event sizes and form the electron edge feature. Moro et al. circumvented this problem by first calibrating the TEPCs using the well-defined proton edge obtained in a low-energy neutron beam on the basis that stopping power data for low-energy protons is much more accurately known than that for low-energy electrons. Using this independent means of calibration they believed the lineal energy associated with the electron edge position would have an uncertainty similar to that of the proton edge at about 3%. For spherical and cylindrical chambers Moro et al. provided the relationships for the electron edge position as a function of simulated diameter TEPCs as

$y_{e,\mathrm{edge}} = 13.9d^{-0.42}$ for spherical counters, with y [keV μm^{-1}] and d [μm] and

$y_{e,\mathrm{edge}} = 15.5d^{-0.42}$ for cylindrical counters

In this work, the electron edge location with the greatest precision was found to be the intercept of the tangent through the inflection point with the event-size axis. This is illustrated in Figure 2 of the Moro paper and reproduced here as Figure 3.11.

Intrinsic calibration using the electron edge becomes extremely important in radiation fields where a proton edge is either not created,

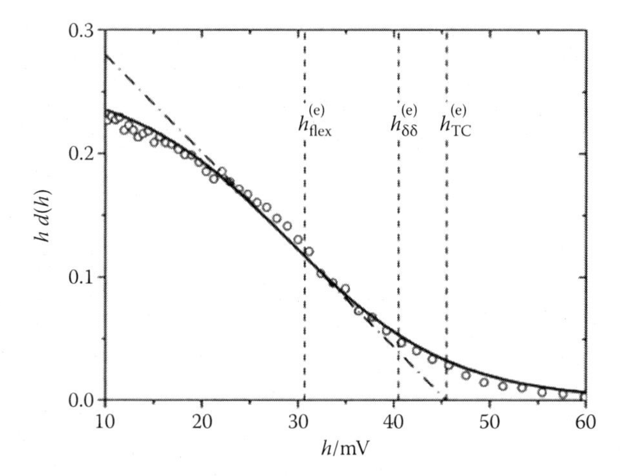

FIGURE 3.11 Identification of the electron edge on the event-size axis by the intercept of the tangent through the edge inflection point, h^{e}_{TC}. This was considered by the authors to be the most precise definition of the 'electron edge'. (Taken from Figure 2 of Moro D. et al., *Radiat. Prot. Dosim.* 166(1/4), 233–237 (2015).)

such as in pure photon fields, or is not easy to identify, as in high-energy neutron and proton fields associated with high-energy accelerators for therapy and physics and space and aviation environments.

3.4.1.4.4 Electron Edge Calibration for Low-Energy Electron Sources

A third elegant calibration method makes use of low-energy electrons from an ^{37}Ar source (Anachova et al., 1994; Chen and Kellerer, 1997). If the range of the electrons is less than the simulated diameter of the cavity the energy of the electrons will be totally absorbed and a peak will appear on the MCA display. ^{37}Ar disintegrates mainly by electron k-capture emitting electrons with energy 2.38, 2.6 and 2.8 keV. The energy imparted is then equal to the energy of the electrons provided the site diameter is greater than the range of the electrons, which is true for site diameters of 1 μm and larger. By dividing the emitted energy by the mean chord length one obtains the lineal energy of these events and by observing the corresponding channel number on the MCA a calibration factor is determined in keV μm^{-1} per channel. ^{37}Ar is obtained by the activation of ^{36}Ar in a thermal neutron beam and so is not a readily obtained product. However, this method of calibration is particularly suited for multi-element TEPCs and wall-less counters as the Ar gas extends throughout the counter volume and a calibration can be obtained for all elements or regions of the counter. The half-life of ^{37}Ar is 35 days, which is long enough for prolonged studies of gas gain stability, but sufficiently short to minimise problems of counter contamination.

A similar methodology can be used with wall-less counters by generating low-energy electrons in the counting gas using low-energy characteristic x-rays of aluminium (1494 eV) or carbon (280 eV) generated by a miniature x-ray set mounted on the side of the detector assembly (Eickel and Booz, 1976).

3.4.1.5 Experimental Uncertainties

Several groups have reported experimental uncertainties. Table 3.4 lists the most common uncertainties. A good review on both electronic as well as non-electronic sources of uncertainties was presented by Kliauga (1990a).

As mentioned earlier, the extrapolation towards zero can be made in many different ways and the fraction of the number of events in this region is about 30% in a ^{60}Co γ beam. The uncertainty in the choice of the extrapolation method was estimated to contribute 6% by Lindborg. Eickel and Booz reported an uncertainty of 20% for the extrapolation region. Varma (1983) reported on different extrapolation methods and derived an uncertainty similar to that by Lindborg.

The calibration process with TEPCs often ignores differences in *W* value between different charged particles. If an instrument has been calibrated with alpha particles and the charged particles dominating the event frequency are protons, as is often the case of neutron irradiation, or electrons in the case of photons a systematic error is introduced. This is so because the basic phenomenon registered by the detector is ion-pair formation. Ionisations can be converted to energy through multiplication

TABLE 3.4 Experimental Uncertainties (in Precision and Accuracy) Reported by Lindborg (1976) for Measurements in a ^{60}Co γ Beam

	$\bar{y}$ (%)	$\bar{y}_D$ (%)
Extrapolation region	6	1
Limited number of events	1	1
Energy calibration	5	5
Pile up	1	1
Gas gain	3	3
Pressure	1	1
Mean chord length	1	
Overall uncertainty (Root mean square)	8	6

with W/e. If the same number of ionisations are created by electrons, N_e, and alpha particles, N_α, ($N_e=N_\alpha$) the imparted energy by the two particles will not be the same because the W/e are not the same. Thus $N_e\,(W/e)_e<N_\alpha\,(W/e)_\alpha$. In precision work, one should try to correct for this by using the proton edge calibration for neutron measurements and the 'electron-edge' or absorption of low-energy electrons, in the case of photon microdosimetry.

3.4.2 Analysis of Multiple Events-Variance and Variance-Covariance Methods

3.4.2.1 Theory

The product of the relative variance, $V_{n,\mathrm{rel},t}$, and the mean value, $\bar{\varepsilon}_t$, of the frequency distribution of energy imparted by multi-events equals the dose mean energy imparted by single events, $\bar{\varepsilon}_{Ds,\,t}$ (Kellerer, 1968):

$$\bar{\varepsilon}_{D,s,t} = V_{n,\,\mathrm{rel},\,t}\bar{\varepsilon}_t \tag{3.25}$$

This theoretical relation was derived in Chapter 2, Equation 2.29 and is the basic relation for the experimental method called the variance method. The dose mean lineal energy, $\bar{y}_{D,t}$, and the dose mean specific energy of single events, $\bar{z}_{D,s,t}$, are both easily calculated, once $\bar{\varepsilon}_{D,s,t}$ is known and

$$\bar{y}_{D,t} = \frac{\bar{\varepsilon}_{D,s,t}}{\bar{\ell}_t} \tag{3.26}$$

$$\bar{z}_{D,s,t} = \frac{\bar{\varepsilon}_{D,s,t}}{m_t} \tag{3.27}$$

The relation Equation 3.25 is independent of the particle fluence as long as $\bar{\varepsilon}_t$ is large enough to ensure that at least one event is present in each sample. This is realised from the following: In a measurement the tissue volume of unit density is simulated by a gas volume, which is much larger. The relative variance of a multi-event distribution, $V_{n,\mathrm{rel}}$, decreases in proportion to the number of events (Equation 2.28 in Chapter 2). At a given fluence, ϕ, the number of events is larger in the detector gas volume than in the microscopic tissue volume and this number is proportional to the cross sectional area of the volume (Equation 3.6). The relative variances observed in the detector and in the tissue volume are thus related as

$$V_{n,\text{rel},\text{det}} = V_{n,\text{rel},t}\frac{\Phi A_t}{\Phi A_{\text{det}}} = V_{n,\text{rel},t}\frac{r_t^2}{r_{\text{det}}^2} \tag{3.28}$$

Equations 3.12, 3.25, and 3.28 give

$$\begin{aligned}\bar{\varepsilon}_{D,s,t} &= V_{n,\text{rel},t}\bar{\varepsilon}_t = V_{n,\text{rel},\text{det}}\frac{r_{\text{det}}^2}{r_t^2}\bar{\varepsilon}_t = V_{n,\text{rel},\text{det}}\frac{r_{\text{det}}^2}{r_t^2}\bar{\varepsilon}_{\text{det}}\frac{r_t^2}{r_{\text{det}}^2}\\ &= V_{n,\text{rel},\text{det}}\bar{\varepsilon}_{\text{det}}\end{aligned} \tag{3.29}$$

A request for variance measurements is that at least one event is depositing energy in each measurement interval. This is seldom the situation for tissue volumes of a few micrometres and at dose rates in the radiation protection field. However, a gas volume simulating those tissue volumes will easily fulfil this condition. The needed scaling factor can be derived from Equation 3.28.

In practice a measurement consists of repeated integrations of ε_i during equally long time intervals and from this series $V_{n,\text{rel},\text{det}}$ and $\bar{\varepsilon}_{\text{det}}$ are calculated. Equation 2.27 in Chapter 2 is equivalent to

$$V_{n,\text{rel},\text{det}} = \frac{\overline{\varepsilon_{\text{det}}^2}}{\bar{\varepsilon}_{\text{det}}^2} - 1 \tag{3.30}$$

With I equal to the total number of measurements in a series

$$\bar{\varepsilon}_{\text{det}} = \frac{1}{I}\sum_{i=1}^{I}\varepsilon_i \tag{3.31}$$

and

$$\overline{\varepsilon_{\text{det}}^2} = \frac{1}{I}\sum_{i=1}^{I}\varepsilon_i^2 \tag{3.32}$$

it follows

$$\bar{\varepsilon}_{D,s} = \left[\frac{\dfrac{\sum_1^I \varepsilon_i^2}{I}}{\left(\dfrac{\sum_1^I \varepsilon_i}{I}\right)^2} - 1\right]\bar{\varepsilon}_{\text{det}} \tag{3.33}$$

A typical value of I is 2000. If the integration is 0.1 s long the measurement will take about 5 minutes to perform.

The variance method is limited to beams in which the variance of the dose rate is dominated by the variance of the energy imparted. If the output of the beam has a non-negligible variance, as is often the case in accelerator and x-ray beams, the variance method may fail. The VC method was derived to overcome this problem (Kellerer, 1996a,b; Kellerer and Rossi, 1984). In this technique two detectors, A and B, are used simultaneously and the relative variance of the energy imparted in each detector as well as the relative covariance of the energy imparted in the two detectors, $C_{\mathrm{rel}}(\varepsilon_{\mathrm{A}}, \varepsilon_{\mathrm{B}})$, are determined. The relation for detector A is

$$\bar{\varepsilon}_{D,s,\mathrm{A}} = \left[V_{n,\mathrm{rel}}(\varepsilon_{\mathrm{A}}) - C_{\mathrm{rel}}(\varepsilon_{\mathrm{A}},\varepsilon_{\mathrm{B}})\right]\bar{\varepsilon}_{\mathrm{A}} \tag{3.34}$$

The covariance is the mean correlated deviation of two random variables from their expectation values and is

$$C(\varepsilon_{\mathrm{A}}\varepsilon_{\mathrm{B}}) = \frac{\overline{(\varepsilon_{\mathrm{A}} - \bar{\varepsilon}_{\mathrm{A}})(\varepsilon_{\mathrm{B}} - \bar{\varepsilon}_{\mathrm{B}})}}{\bar{\varepsilon}_{\mathrm{A}}\bar{\varepsilon}_{\mathrm{B}}} = \frac{\overline{\varepsilon_{\mathrm{A}}\varepsilon_{\mathrm{B}}}}{\bar{\varepsilon}_{\mathrm{A}}\bar{\varepsilon}_{\mathrm{B}}} - 1 \tag{3.35}$$

In a measurement series of I equally long time intervals the energy imparted in the two detectors are $\varepsilon_{\mathrm{A},\,i}$ and $\varepsilon_{\mathrm{B},\,i}$. The relative variance of detector A is

$$V_{n,\mathrm{rel}}(\varepsilon_{\mathrm{A}}) = \frac{1}{I}\sum_{i=1}^{I}\frac{\varepsilon_i^2}{\bar{\varepsilon}_{\mathrm{A},\,i}^2} - 1 \tag{3.36}$$

The relative covariance is

$$C_{\mathrm{rel}}(\varepsilon_{\mathrm{A}}\varepsilon_{\mathrm{B}}) = \frac{1}{I}\sum_{i=1}^{I}\frac{\varepsilon_{\mathrm{A},\,i}\varepsilon_{\mathrm{B},\,i}}{\bar{\varepsilon}_{\mathrm{A}}\bar{\varepsilon}_{\mathrm{B}}} - 1 \tag{3.37}$$

The dose mean energy imparted measured by detector A is thus

$$\bar{\varepsilon}_{D,\,s,\,\mathrm{A}} = \frac{1}{I}\sum_{i=1}^{I}\varepsilon_{\mathrm{A},i}^2/\bar{\varepsilon}_{\mathrm{A}} - \frac{1}{I}\sum_{i=1}^{I}\varepsilon_{\mathrm{A},i}\varepsilon_{\mathrm{B},i}/\bar{\varepsilon}_{\mathrm{B}} \tag{3.38}$$

$\bar{\varepsilon}_{D,s,B}$ is calculated in the same way.

The preceding equations are applicable when the ratio of the dose rate measured by the two detectors can be assumed to be constant. This may not always be true. The detector system may be moving in a non-homogeneous radiation field or there may be movement of scattering material near the detectors. In this situation and if the ratio of the dose rates observed by the two detectors is only slowly changing with time a further generalisation of the method has been suggested and successfully tested by Kellerer (1996b,c). Here

$$\bar{\varepsilon}_{D,sA} = \frac{1}{2} \times \left\{ \frac{\sum_{i=1}^{I-1} (\varepsilon_{A,i} - \varepsilon_{A,i+1})^2}{\sum_{i=1}^{I-1} \varepsilon_{A,i}} + \frac{\sum_{i=1}^{I-1} (\varepsilon_{A,i} - \varepsilon_{B,i})^2}{\sum_{i=1}^{I-1} \varepsilon_{B,i}} - \frac{\sum_{i=1}^{I-1} (\varepsilon_{B,i} - \varepsilon_{A,i+1})^2}{\sum_{i=1}^{I-1} \varepsilon_{B,i}} \right\} \quad (3.39)$$

$$\bar{\varepsilon}_{D,sB} = \frac{1}{2} \times \left\{ \frac{\sum_{i=1}^{I-1} (\varepsilon_{B,i} - \varepsilon_{B,i+1})^2}{\sum_{i=1}^{I-1} \varepsilon_{B,i}} + \frac{\sum_{i=1}^{I-1} (\varepsilon_{A,i} - \varepsilon_{B,i})^2}{\sum_{i=1}^{I-1} \varepsilon_{A,i}} - \frac{\sum_{i=1}^{I-1} (\varepsilon_{A,i} - \varepsilon_{B,i+1})^2}{\sum_{i=1}^{I-1} \varepsilon_{A,i}} \right\} \quad (3.40)$$

Higher moments of the energy imparted can also be derived (Kellerer, 1996a).

If the mean number of events during a sampling interval is known it is also possible to determine the frequency mean energy imparted of the single event distribution, $\bar{\varepsilon}_S$. In principle for photons a GM counter at the position of the proportional counter could be used for this purpose. If the mean number of events/sampling interval is $\bar{n}$ and the mean energy imparted/sampling interval is, $\bar{\varepsilon}$, determined from VC measurements, then $\bar{\varepsilon}_S = \bar{\varepsilon}/\bar{n}$.

3.4.2.2 Variance Measurement Techniques

The possibility to experimentally determine $\bar{\varepsilon}_{D,s}$ from Equation 3.25 was first realised by Bengtsson (1970). To achieve multiple events the dose is accumulated for a fixed time and from a series of such integrations the variance is calculated. Both the time and the dose per integration have to be measured with high accuracy. Electrometers are suitable for this kind of measurement and are applicable both in radiation fields of high dose rates as well as in radiation fields from pulsed beams, situations in which single event measurements usually are complicated or impossible to carry out (Mayer et al., 2004). Electrometers have been used either with a capacitor or a resistor in the feedback circuit. In any case, it is important that the resistor and the capacitor are of high quality. Electric circuits are shown in

Figure 3.12a and b. The capacitive method will allow shorter integration times and is expected to have lower noise level as with resistors in the feedback circuit thermal noise usually becomes large. Time intervals ranging from 1 ms to 0.2 s have been reported together with capacitive feedback, while time intervals in the range of 0.1 s to 0.6 s were used with a resistive feedback circuit (10^{11} Ω). The voltage observed at the output of the electrometer needs to be digitised with high resolution and the data transferred to a computer for analysis. Technical details have been reported by a number of workers (Bengtsson, 1970; Braby, 2015; Breckow et al., 1988; Grindborg et al., 1995; Lillhök et al., 2007c; Lindborg et al., 1989). The PHA technique was applied to VC measurements in a pulsed beam by Kliauga et al. (1986). The accelerator pulse was made to trigger the charge sensitive amplifier that records the charge during the short accelerator pulse. After amplification the charges were digitised in an ADC and were registered in the MCA working in list mode or multichannel scaler mode; that is, pulses are stored one after the other in the channels. The variance and covariance are then calculated from the stored data. The linearity of the preamplifiers is critical and must be checked.

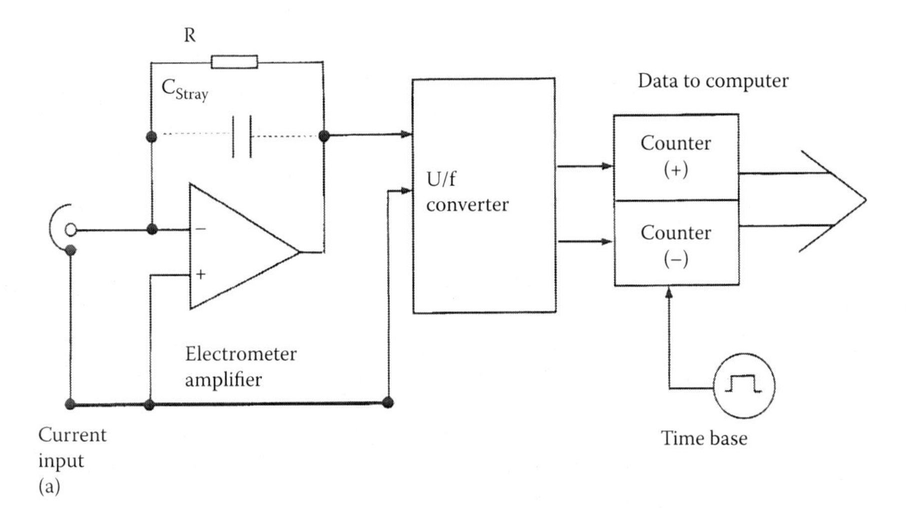

FIGURE 3.12 (a) Block diagram of an electronic circuit for variance measurements based on an electrometer using resistive (R) feedback. (From Grindborg J.-E. et al., *Radiat. Prot. Dosim.* 61, 193–198 (1995).) The output voltage is converted to frequency by a volt/frequency-converter (U/f) and a counter registers the pulses in preset time intervals. Two converters are used to allow measurements of both positive and negative currents. *(Continued)*

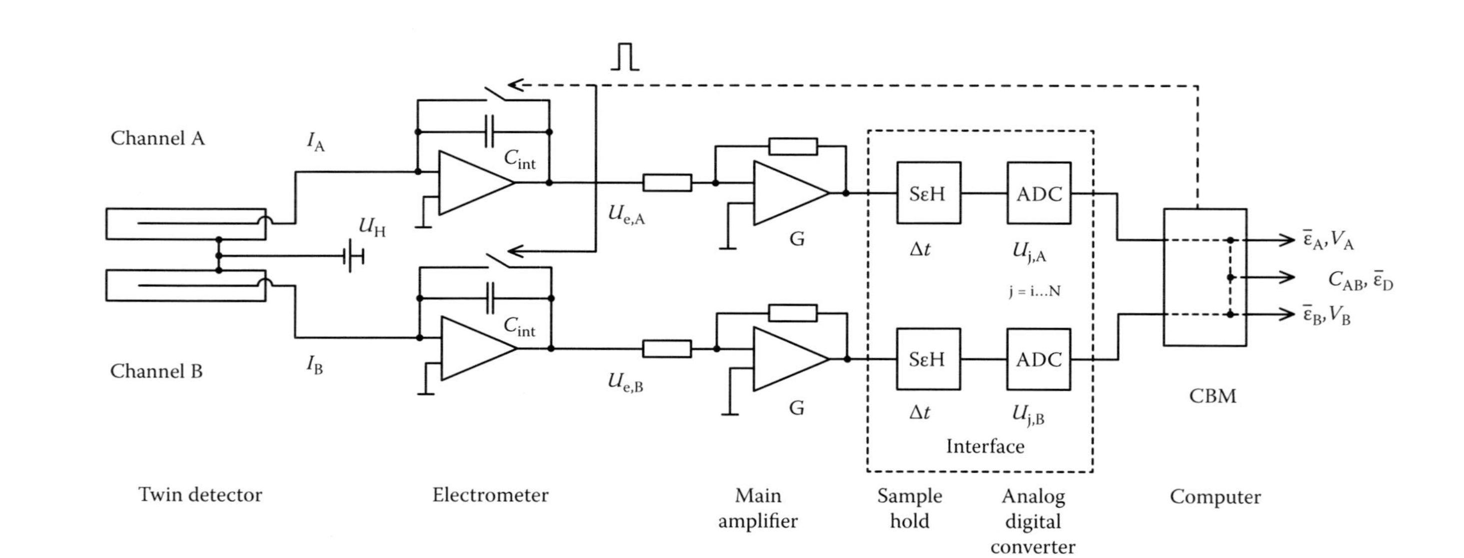

FIGURE 3.12 (CONTINUED) (b) Block diagram of the electronic circuit for VC measurements using electrometers with capacitive (C_{int}) feedback. (From Breckow J. et al., *Radiat. Environ. Biophys.* 27, 247–259 (1988).)

In electrometer measurements the important quantity is the electric charge created through ionisations in the detector gas. To make sure that all ions or electrons created are collected it becomes necessary to determine the electric current as a function of the applied potential and determine the saturation plateau. A correction for ion recombination is seldom necessary but its importance should be judged on a case-by-case basis. Such a measurement is needed for each gas pressure used. If gas multiplication is needed it has usually to be experimentally determined from measurements in radiation beams of different intensities.

To arrive at the mean energy imparted a conversion from electric charge is required. The mean energy expended in a gas per ion pair formed, W (ICRU, 1979), is defined for this purpose and the relationship between the measured electric charge, Q, and $\bar{\varepsilon}$ is

$$\bar{\varepsilon} = QW/e \tag{3.41}$$

and e is the electric charge of the electron (1.602×10^{19} C). If gas multiplication is used and the amplification factor is G, the relation is

$$\varepsilon = \frac{Q_e}{G} W/e \tag{3.42}$$

W is not a constant but varies with the energy of the charged particle. It is different for different charged particles and varies also with the gas. At very low energies (usually below 100 eV) the value will increase rapidly (Grosswendt, 2002; ICRU, 1979). According to the definition of W the complete number of electric charges of one sign along a particle track have to be collected. This is seldom possible, as in the experimental situation the cavity has a limited extension and part of the particle energy may be dissipated in the wall and some of the secondary electrons created in the gas may also continue into the wall. However, if charged particle equilibrium exists, different parts of a given particle track will ionise the gas and W should be an acceptable approximation.

This situation is presumed to hold at least for large volumes. In very small volumes (nanometre range) the ionisation is mainly by very low energy electrons and the energy dependence of the W value may need to be considered (Section 3.5). When the radiation field is a mixture of different charged particles the dose fractions of the different charged particles have to be estimated and a dose weighted average W value should be applied.

In the VC method two simultaneously measuring detectors with electrometers are needed. The detectors do not necessarily both have to be TEPCs. A dosemeter of good quality could be used as one detector. It is important that both detectors start measuring simultaneously. The time constants of the measuring system need to be the same within a few milliseconds. Another requirement is that the two detectors are responding independently to the absorbed dose. If, for instance, a high-LET charged particle is passing through both detectors A and B an increase of ε may occur in both chambers such that $\varepsilon_A = k\varepsilon_B$ where k is a constant factor, an increase in dose will appear in both at the same time. This will be interpreted as a covariance and the evaluated dose mean value will become too low. The positioning of the detectors therefore requires some attention [an example has been given by Magrin et al. (2000)]. Another important point is that the probability of having no events in a sampling interval must be small.

To decide whether a value of $\bar{\varepsilon}_D$ is of acceptable quality repeated measurements with different experimental settings are needed. Equation 3.25 shows that if the mean energy imparted, $\bar{\varepsilon}$, during the sampling time interval is increased, its relative variance, $V_{\text{rel},\bar{\varepsilon}}$, will decrease, as $\bar{\varepsilon}_{D,s}$ is constant and independent of $\bar{\varepsilon}$. Parameters to change are the sampling time length, the dose rate and the gas multiplication. If $\bar{\varepsilon}_{D,s}$ remains independent of any changes in $\bar{\varepsilon}$ then its value can be expected to be accurately determined.

The measurement system must be checked for leakage and/or background currents before and after a measurement and if necessary a correction made. If a constant leakage current exists, the final value can be corrected as

$$\bar{\varepsilon}_{D,s,A} = \left[\frac{V'_{r,A}}{1-\frac{Q_{lA}}{\bar{Q}'_A}} - \frac{C'_r}{1-\frac{Q_{lB}}{\bar{Q}'_B}}\right] Q'_A W/e \qquad (3.43)$$

Here $\bar{Q}'_A$ and $\bar{Q}'_B$ are the mean electric charges during a measurement series, Q_{lA} and Q_{lB} are the constant leakage charges in the two detectors and $V'_{r,A}$ and C'_r are the measured relative variance and relative covariance (Lindborg et al., 1989). Alternatively, electrometers may be adjusted so that a constant offset current compensates for this current and no correction is needed. Corrections for variances due to background and leakage currents have been reported by Grindborg et al. (1995).

When an ionisation chamber is used, the same result is expected whether ions or electrons are collected. Usually there are minor differences due to design details, and mean values of measurements at both polarities are recommended when high accuracy is needed. When measurements are extended towards smaller simulated volumes, it is useful to check that the electric current corrected for background and secondary electron emission (SEE) is decreasing linearly with pressure.

The SEE is a current of electrons originating from the innermost part of the wall and the central collecting electrode with such low energies that they are unable to ionise the gas (Burlin, 1974). This energy limit is somewhat arbitrarily set to 50 eV. As these electrons are not generated through ionisations in the gas, they should not be included in the current measurement. The SEE current can be determined with the detector fully evacuated and with a voltage of +50 V applied to the collecting electrode. When the detector is irradiated there will be no ionisation in the cavity, as there is no gas. Electrons with energies larger than 50 eV will have sufficient energy to escape the collecting electrode, while electrons with lower energies will be recaptured. As this current is proportional to the irradiated surface, a solid chamber wall will generate a large quantity of these electrons. The current will be much reduced if ions are collected instead of electrons (Forsberg and Lindborg, 1981; ICRU, 1983). It will also be low when a wall-less chamber is used. The SEE current is an important consideration when small volumes are investigated (Section 3.3).

An evaluation of a VC measurement is illustrated with a measurement series in a monoenergetic neutron beam of 19 MeV. Two cylindrical TEPCs 10 cm internal diameter and 10 cm internal height were used. They simulated a mean chord length of 1.95 μm. The gas multiplication factor G was 322. Electrometers with capacitive feedback were employed and the voltage across the feedback capacitor, U, was sampled every 0.3 s. The differences between two consecutive U values, are then calculated and multiplied with the calibration factor for the electrometer and divided by the gas multiplication factor. Figure 3.13 presents the electric charge per sampling interval, 0.3 s, for the whole measurement series. Its mean value, $\bar{Q}_e$, is 35.8 fC and its relative standard deviation $\sigma_{\text{rel}} = 0.179$. The mean energy imparted is then

$$\bar{\varepsilon}_{\text{det}} = \bar{Q}_e W/e = 35.8 \times 10^{-15} \times 30 = 1.07 \times 10^{-12} J$$

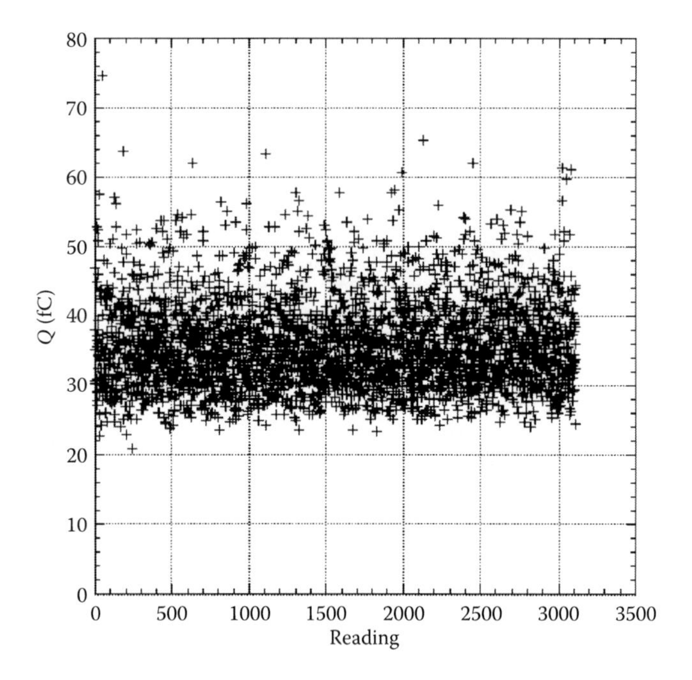

FIGURE 3.13 Results obtained with VC measurements. The figure shows differences between consecutive values of the electric charge. The time interval between two readings is 0.3 s; thus 1000 values correspond to 300 s. (Experimental data from J. E. Lillhök, private correspondence, 2016.)

The covariance was small and the dose mean lineal energy is

$$\bar{y}_D = \frac{\bar{\varepsilon}_{\text{det}}\sigma_{\text{rel}}^2}{\bar{\ell}_s} = 110\,\text{keV}\,\mu\text{m}^{-1} \tag{3.44}$$

The mean specific energy, $\bar{z}_t = \dfrac{\bar{\varepsilon}_{\text{det}}}{m_t} = 51$ Gy where $m_t = 21 \times 10^{-15}$ kg is the mass of the tissue volume with mean chord length 1.95 μm. The mean absorbed dose in the detector gas per integration interval is $D = \dfrac{\bar{\varepsilon}_{\text{det}}}{m_{\text{det}}} =$ 3.4 n Gy where $m_{\text{det}} = 3.1 \times 10^{-5}$ kg is the mass of the detector gas. The dose rate is then 0.4 mGy h^{-1}.

A multi-event distribution is created by dividing the range of specific energy into a suitable number of equally large intervals. In this example, an interval width of 1 Gy turned out to give sufficiently good statistics. The mean number of specific energy values in each interval is – after

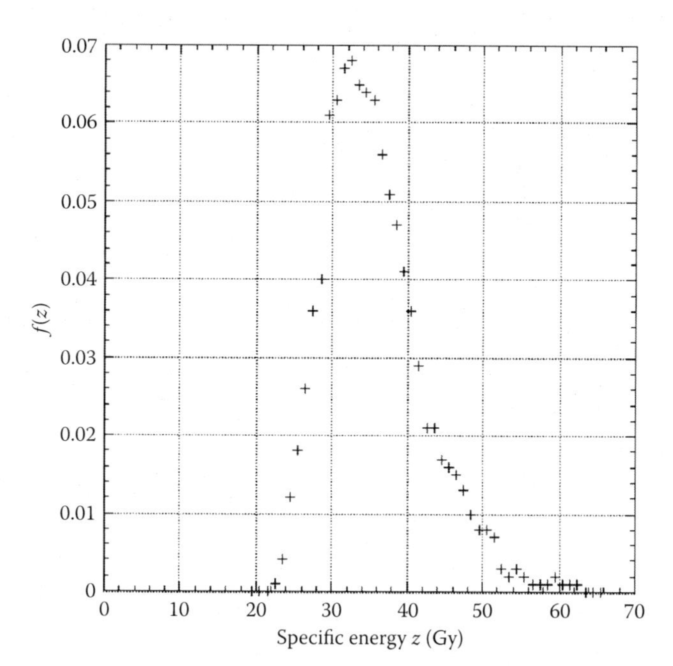

FIGURE 3.14 The multi-event frequency distribution of the specific energy observed in a 19 MeV neutron beam with a cylindrical TEPC simulating a mean chord length of 2 μm. The distribution is derived from the results in Figure 3.13. Each data point corresponds to the number of multi-events in 1-Gy intervals. The mean specific energy, $\bar{z}$ = 35.8 Gy.

normalisation to unity – plotted as function of the specific energy in Figure 3.14.

3.4.2.3 Calibration for Lineal Energy

Calibration for lineal energy in the variance method is quite different from the methods described above for single event distribution measurements. The variance and the VC techniques may be regarded as an independent means for the determination of $\bar{y}_D$, which is obtained from the following relation already described

$$\bar{y}_D = \frac{Q_e W V_{n,\mathrm{rel}}}{e\bar{l}_t} \tag{3.45}$$

Any penetrating radiation field can be used for calibration if its W is known. A calibrated electrometer is the only further need. The procedure for electric charge described previously has to be followed. $\bar{\ell}_t$ is determined in the same way as in single event analysis. The choice of correct

W value is important. *W* values have been published by ICRU (1979). For propane-based TE gas such values are less easy to find (Bronic, 1997; Grosswendt, 2002).

3.4.2.4 Experimental Uncertainties

The measured variance as well as the measured mean energy imparted is affected by several experimental parameters (influence quantities) (Forsberg and Lindborg, 1981; Grindborg et al., 1995; Kliauga et al., 1986; Lindborg et al., 1989). Their influence on the result has to be small. With the nomenclature of ISO, uncertainties may be called type A and type B (JCGM, 2008). The former are traditional statistical uncertainties, while type B are usually estimated by other means (earlier called systematic uncertainties). Table 3.5 shows the combined uncertainties for the influence factors listed in the text that follows. A standard uncertainty (corresponding to a confidence level of 95%) is shown.

3.4.2.4.1 Type A Uncertainties

1. The number of integrations, *I*, in a measurement series affects the uncertainty of the variance and the 95% confidence level is approximately $1.37/\sqrt{I}$ (Natrella, 1966). If $I = 2000$, $\sigma_{rel,var} = 0.031$ or 3.1%. The experimental uncertainty in the mean value of both detectors as function of the number of integrations, *I*, is in good agreement with this expected theoretical value. From measurements with the VC method the uncertainty in the relative covariance term was reported

TABLE 3.5 Influence Factors as Reported for the VC Method by Different Authors

Influence Factor	Lillhök et al. (2007c) (%)	Grindborg et al. (1995) (%)	Lindborg et al. (1989) (%)
W value	2		
Detector diameter	6	4	
Electrometer	4	3	
Pressure in the detector	4		
Variance	2	4	
Standard uncertainty	9	6	6

Note: Values from measurements in a neutron therapy beam and mean chord lengths above 50 nm (Lillhök et al., 2007c), an x-ray-beam and a ^{60}Co γ beam (Grindborg et al., 1995) and a neutron beam of 5.7 MeV and mean chord lengths above 20 nm (Lindborg et al., 1989). The estimated 95% confidence level is presented on the bottom line.

to be equal to that of the relative variance and the two have to be added as the sum of the squares (Lindborg et al., 1989). Usually this uncertainty dominates the type A uncertainty for volumes above 50 nm.

2. A variance contribution due to electronic noise exists and a standard deviation of a few femtoamperes has been reported (Magrin et al., 2000). To achieve low electronic noise, low-noise transistors at the input of the preamplifier are of importance. The electrometers may be split into a preamplifier part separated from the rest of the electrometer. The preamplifier can then be connected to the chamber directly or with a short cable, which reduces input capacitance and noise. This arrangement is seldom a problem in radiation protection measurements, where the dose rate is low. In an environment around a therapy beam the dose from stray radiation may be high and may induce currents through ionisation of the air inside the preamplifier housing. Stowing the preamplifier inside a radiation shield outside the primary beam or even evacuating the gas inside it are ways of improving the situation.

3. Electrometers measure currents in the femtoampere range with quite good accuracy and above 0.1 pA the accuracy is usually $<10^{-3}$. However, if that current is observed for 0.1 s the electric charge is only 10 fA and the number of electrons collected is $n_e = (10^{-14}/1.6 \times 10^{-19}) = 6250$. The relative standard deviation due to the limited number of electrons is $\sqrt{62,500} = 250$ or 0.4%. A high dose rate, a large detector, gas multiplication or a combination of the three will keep this contribution small. In practice and for other reasons as well a current of at least 0.5 pA is usually required for successful measurements.

3.4.2.4.2 Type B Uncertainties

1. The diameter of the detector volume can be determined from a drawing of the detector or from an x-ray image of it. As already mentioned, Kliauga has reported a shrinkage of the A-150 shells with time. If a wall-less counter is used its wall is quite fragile and a precise measure of its diameter may be difficult to obtain.

2. Reading of a pressure meter. Although pressure meters are usually calibrated at a standards laboratory the pressure may be difficult to

measure in the volume when gas flow is applied. This is particularly true for small counter volumes or high gas flow.

3. The uncertainty in the *W* value has been given by the ICRU. When the simulated volume gets smaller a differential *W* value should be considered and when a mixture of charged particles is interacting in the gas a mean *W* value should be used. These factors increase the uncertainty in the *W* value and have to be evaluated for the specific field being investigated.

4. The SEE phenomenon will add to the uncertainty both in the current and the variance unless a correction is made.

5. The environment temperature may affect the gas multiplication, if different from the calibration temperature.

3.4.2.5 Methods for Deriving Dose Fractions

The variance method yields the dose mean energy imparted in single events for the beam under investigation but in radiation protection; for instance, there may be a need to separate the dose component from alphas from that of protons and electrons as these dose fractions will have different weighting factors. Methods to manage this situation are covered in Chapter 5. A similar situation exists in radiation therapy with protons and ions although the differences between the weighting factors are smaller. Methods are covered in Chapter 4.

3.4.3 Comparison of the Variance Method and the PHA Method

The two principal methods in experimental microdosimetry, variance method and PHA, have been compared in a few specific tests. Early intercomparisons have been reported for a neutron beam (Bengtsson, 1970) and a ^{60}Co γ beam (Bengtsson and Lindborg, 1974). An agreement within about 10% was reported. A wall-less detector (Figure 3.15), which allowed for both the VC method as well as PHA method was constructed by Rossi and has been used in two different comparisons. Methane-based TE gas was used with the VC method, while propane-based TE gas was used when the PHA method was applied. The first comparison was made in a neutron beam of 5.7 MeV and three simulated volumes between 0.3 μm and 3 μm were investigated. The difference was on average 6% (Lindborg et al., 1989). The same detectors and similar conditions but in a neutron beam of 15 MeV have been reported (Goldhagen et al., 1990). An

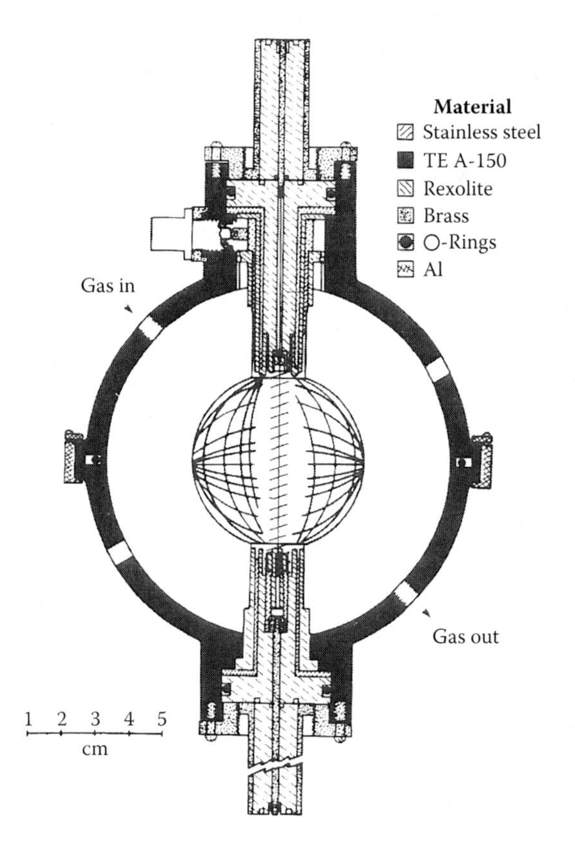

FIGURE 3.15 A cross section of a wall-less spherical proportional chamber designed for both the variance and the PHA method of measurement. The detector was designed by H. H. Rossi and manufactured at Columbia University. To allow for electrometer measurements the central electrode is furnished with an electric guard. (From Lindborg L. et al., *Radiat. Environ. Biophys.* 28, 251–263 (1989).)

excellent agreement was reported between the two methods and volumes between 0.3 μm and 5 μm. Comparisons of results obtained with the two methods are also described in Chapter 5.

3.5 LIMITATIONS IN SIMULATIONS OF VOLUMES IN THE NANOMETRE RANGE

When gas multiplication is used, the resolution will gradually decrease with decreasing gas pressure. To keep the pressure sufficiently high a few groups have developed miniature counters (Kliauga, 1990b; Moro et al., 2009). They constructed miniature cylindrical counters with cavities of 0.5 × 0.5 mm and 0.9 × 0.9 mm respectively. Kliauga reported lineal energy distributions down to 5 nm, while Colautti reported distributions down to 50 nm. Cesari et al. (2001) reported results from measurements in photon and neutron beams down to 35 and 25 nm respectively. Another approach was reported by Anachova et al. (1997). Here a 60 mm long cylindrical TEPC only 3 mm in diameter was used and single-event distributions from measurements in a neutron beam down to 50 nm were reported. Chen et al. (1990) reported measurements in a ^{137}Cs beam down to 67 nm. The electric charge collected through SEE from the wall of a proportional counter will affect the PHA method of measurement as well as that of the VC method. Procedures for determining the magnitude of the SEE current for the VC method were given in Section 3.4.2.2. The effects of SEE are not generally discussed in reports on PHA microdosimetry but they can be expected to become more important as the simulated diameter and gas pressure are reduced. As SEE will be proportional to the surface area of a counter wall some mitigation of this effect will be gained by using miniature counters as is mostly the case in nanometre dosimetry.

The VC method was early applied to measurements in the nanometre range using ionisation chambers and TEPCs operating without gas gain. To compensate for the weaker signal at lower pressures, measurements are made in beams of high dose rates typically such as those used in radiation therapy. Detectors a few centimetres in diameter have been used. Wall-less detectors are also preferred as wall effects as well as the SEE current will be smaller. The SEE current is also reduced if ions are collected instead of electrons. The electrometers shall have a low noise and preferably placed as close as possible to the detector. To illustrate what limits the possibility to simulate small volumes with the variance-covariance techniques, results measured with two different kinds of detectors will be examined in some detail.

A wall-less TEPC for VC measurements was developed at Columbia University (Lindborg et al., 1989) (Figure 3.15). It was furnished with electric guard not just around the centre electrode but also around the helix to allow the electrometers to be connected to the helix as an alternative to the central electrode. The diameter of the grid was 5.1 cm. The outer shell was made of A-150 and had a diameter of 10.2 cm. The detector could operate not only as a proportional counter but also as an ionisation chamber. Measurements with this detector have been reported for both a ^{60}Co γ beam as well as different neutron beams. The second type of detector was commercially available ionisation chambers (Exradin A3 and A4). The detectors are spherical walled ionisation chambers

38 mm and 57 mm in diameter made of air equivalent plastic (C-552) and during the measurements filled with air. The detectors are furnished with electric guards and are well suited for electric current precision measurements. They were used for measurements in the same ^{60}Co γ beam as the TEPCs, but also in an x-ray beam (100 kV, added filter 3.14 mm Al and HVL 0.141 mm Cu). The measured values were converted to $\overline{y}_D$ in tissue (Grindborg and Olko, 1997).

Technically the procedure at each pressure (or simulated tissue volume) consisted of establishing the ionisation saturation current and when necessary corrected for recombination. The variances from SEE current, radiation induced leakage currents, cables and connectors (including electronic noise) were determined during irradiation of the gas-evacuated detectors. These variances are independent of the gas pressure and will increase in importance with decreasing pressure. The different variance sources for the A3 ionisation chamber in the two beams are plotted in Figure 3.16 (Grindborg et al., 1995). At about 15 nm the two groups of variances are comparable in magnitude. Determinations of $\bar{y}_D$ below 15 nm with those detectors require carefully determined corrections and results below this value will have less accuracy.

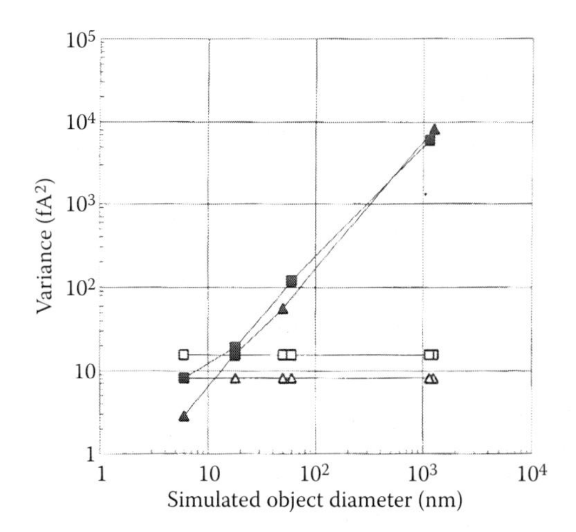

FIGURE 3.16 The measured microdosimetric variance for different simulated diameters compared to the measured charge variance with evacuated chambers (background variance). Results from measurements in a ^{60}Co γ beam (filled squares) and an x-ray beam (filled triangles) with dose rates 9 mGy s^{-1} and 6 mGy s^{-1} respectively. The unfilled symbols are the results from the background measurements. The background variance has been subtracted from the measured variance.

The measured $\bar{y}_D$ values in the low-LET beams were also compared to Monte Carlo–track-structure calculated $\bar{y}_D$ values using the electron transport code MOCA-8. The secondary electron spectra were calculated using the PHOEl-2 code. The calculations were made in water vapour and only ionisations were counted and converted to energy imparted by multiplication with *W*. In this way, the calculated values were directly comparable to the measured results. (The detector itself was not simulated in the calculation.) Figure 3.17 shows both the measured and the calculated $\bar{y}_D$ at different mean chord lengths in the ^{60}Co γ beam with both the wall-less TEPC and the ionisation chambers. The results obtained with the wall-

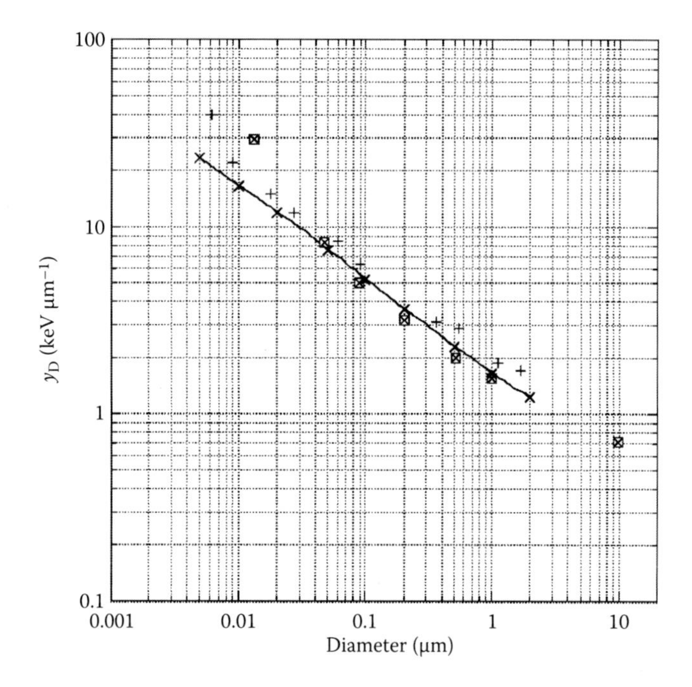

FIGURE 3.17 Results from measurements with the VC method and two wall-less TEPCs of the type shown in Figure 3.15 positioned inside a phantom made of A-150 TE plastic (crosses in squares). (From Lillhök J. E. *The microdosimetric variance-covariance method used for beam quality characterization in radiation protection and radiation therapy*. Thesis, Medical Radiation Physics, Stockholm University and Karolinska Institutet (2007).) Results of measurements with a walled spherical air equivalent ionisation chamber (+) and calculations made with the Monte Carlo track structure code MOCA-8 are shown for the same ^{60}Co γ beam. (From Grindborg J.-E. and Olko P. A comparison of measured and calculated-values in the nanometre region for photon beams. In D. T. Goodhead, P. O'Neill and H. G. Menzel (eds.), *Microdosimetry: An interdisciplinary approach. Twelfth symposium on microdosimetry* (pp. 387–390) (1997).)

less counters are in good agreement with the calculated results down to at least 50 nm. Below that, a difference appears and increases with decreasing diameter. The agreement between measured and calculated $\bar{y}_D$ values confirms that measurements of the dose mean of ionisation distributions with the VC method can be made reliably down to at least 50 nm site diameters. The measured $\bar{y}_D$ obtained with the air-equivalent ionisation chamber were reported to be systematically 20% larger in the range 9 nm to 2 μm. Part of this is due to wall effects (Section 3.3.1.2). If the measured results are multiplied by 0.8, calculated and measured results at 10-nm site diameters and above are not significantly different, while at 6 nm the experimentally determined $\bar{y}_D$ is 1.5 times larger. These two examples indicate that somewhere in the range of 10 to 50-nm site diameters there is a practical limit to what volumes can be simulated with presently available detectors and techniques.

The measured results obtained at very small volumes are affected by ignoring the increase in W with decreasing particle energy as well as ignoring the escape of electrons from the volume under observation. The increase in W is connected with an increased fraction of energy transferred in excitations, which above all affects energy transferred by electrons with energies below 100 eV (Nikjoo et al., 2011). The escape of electrons from the volume can be corrected for if a differential W (notation w) is applied. The expected influence has been calculated by Brenner and Zaider (1984) as well as Amols et al. (1990). In the first paper $\bar{y}$ and $\bar{y}_D$ were calculated in water vapour for protons of 1.5 MeV when only ionisations multiplied with the differential w value were considered. They also calculated the energy imparted in the volume when both ionisations and excitations were included. For 1.5-MeV protons and a spherical volume with 10 nm diameter $\bar{y}$ calculated from ionisations only was a factor 2 above that when all energy transfers were included, while $\bar{y}_D$ was overestimated by almost 20%. At 100 nm the differences were less than 10% for both mean values. Amols et al. (1990) reported a similar study for 200 keV electrons using Monte Carlo track segment calculations for water vapour. The calculated number of ionisations was multiplied with W to obtain the lineal energy distribution from ionisations only and this distribution was compared to the distribution when all energy transfers were included. At a simulated volume with the diameter 10 nm $\bar{y}$ calculated from the ionisations only was a factor two larger than $\bar{y}$ calculated from the distribution when all energy transfers were included. For $\bar{y}_D$ the increase was about 20%. While an influence on $\bar{y}$ was observed already at 1 μm, $\bar{y}_D$

was influenced only below 0.1 μm. Thus $\bar{y}_D$ measured at a diameter of 10 nm with the VC method and with constant W values can be expected to be about 20% larger than the true $\bar{y}_D$. Bigildeev and Lappa (1994) have derived theoretical methods to correct measured mean energy imparted, $\bar{\varepsilon}$, as well as dose mean energy imparted, $\bar{\varepsilon}_D$, determined from ionisation measurements when a constant W has been used. Grindborg et al. (1995) estimated this correction factor to be 0.9 for $\bar{\varepsilon}_D$ determined with the VC method at 9 nm and a ^{60}Co γ beam.

Another question of interest when moving into the nanometre range is whether the simulation principle still holds. Grosswendt (2002) investigated whether a tissue equivalent gas volume could simulate a biological target or at least a target of liquid water when the dimension is expressed in terms of $(\bar{\ell}\varrho)_{\text{TE gas}} = (\bar{\ell}\varrho)_{\text{water}}$, the basic relation for calculation of simulated mean chord lengths. The investigation was made with Monte Carlo track structure simulations of electrons and alpha particles in liquid water and propane-based TE gas. The criterion compared in the two materials was the particle range that includes 95% of the total number of ionisations. The calculations showed that for a 100-eV electron a target thickness of 0.45 μg cm^{-2} in liquid water and 0.28 μg cm^{-2} in the TE gas is needed (corresponding to 4.5 nm and 2.8 nm respectively at unit density) to fulfil the criteria. For a 1-keV electron the corresponding target thickness is 4.5 μg cm^{-2} and 4.0 μg cm^{-2} respectively corresponding to 45 nm and 40 nm. (The ranges in liquid water can be compared to the results in Figure 3.9.) These calculations suggest the relationship between tissue and simulated tissue volumes in particular below 10 nm is complex and nontrivial. Grosswendt found that the mean number of ionisations in the two materials would be close if the value measured in the gas is corrected for the difference in mean free path length with respect to ionisations between water and gas. Support for this conclusion was found when ionisation yields of alpha particles of 5 MeV and electrons from 20 eV to 10 keV were calculated. Methods for measurements in the nanometre range are dealt with in the next section.

3.6 IONISATION CLUSTER DISTRIBUTIONS MEASUREMENTS IN NANOMETRE VOLUMES

A few groups have applied an alternative approach for measurements in the nanometre range (De Nardo et al., 2002; Pszona et al., 2000; Shchemelinin et al., 1999). Common to these studies is that the results of the measurement process are *ionisation* distributions. In this way, the problem of uncertainty connected with energy calibration in small volumes is overcome. In the technique used by Shchemelinin and Pszona ions are collected while in the technique used by De Nardo electrons are collected. Common also is that the charged particles created are collected outside the simulated volume where they are multiplied in either an electron multiplier (Shchemelinin and Pszona) or a special proportional counter (De Nardo). In this way, the problem of increasing variance from the gas multiplication inside the sensitive volume is avoided. The simulated volumes are wall-less (Shchemelinin and De Nardo) and defined by a sophisticated electrode arrangement. In the methods of Shchemelinin and De Nardo the pressure of the gas in the detector is kept constant, while in the experimental arrangement of Pszona, called a jet counter, gas is injected as a short pulse into the sensitive volume. In the experiment of Shchemelinin volumes of diameter from sub-nanometre up to several nanometres are possible, in the De Nardo counter volumes of about 20 nm diameter are reported and in the jet counter technique it is possible to measure simulated volumes with diameters from 0.2 to 13 nm. The detector used by De Nardo is designed for studying ionisation distributions at different distances from a primary ion track. Large corrections for the detection efficiency are reported for the methods (De Nardo et al., 2011). A more detailed review of the three methods has been presented by Schuhmacher and Dangendorf (2002). Cloud chamber techniques have been used for the same purpose and have been reviewed by Laczkó (2006). The formalism used for measurements of cluster distributions in the nanometre range as well as the challenges in such measurements has been reviewed by Palmans et al. (2015). Whether ionisations rather than energy imparted is better correlated with complex clustered damages in DNA is an open question (Garty et al., 2010).

3.7 GAS FLOW SYSTEM AND GAS GAIN CONTROL

In order to simulate tissue equivalent volumes the gas pressure has to be reduced (Table 2.2) unless the detector is very small. All material including the material A-150 adsorbs to some extent molecules from the gas. When the pressure is reduced the adsorbed molecules will gradually outgas from the wall changing the composition of the counting gas. As gas multiplication and resolution are dependent on the gas composition in spite of the precautions taken described above a continuous gas flow through the detector is often used to improve the constancy of the gas property for high-accuracy measurements.

A typical gas flow system is schematically shown in Figure 3.18. The gas bottle is connected to the gas flow system through a valve and an automatic regulating valve; the latter is connected to a nearby pressure meter, which feeds back information on the pressure to the valve. The gas then passes through the detector and passes one more pressure meter after which there is another fine needle valve before the pump. Usually it is practical to have both a roughing rotary pump as well as a diffusion or turbo-molecular pump. The latter is used only to speed up the evacuation of the counter and gas flow system. The tubes connecting the detector to the gas flow system may have quite small diameters. For that reason and during measurements the gas flow cannot be too high as this will create a pressure gradient across the detector and introduce an uncertainty in the volume simulated. The two pressure gauges on both sides of the detector are used to observe this. To make the evacuation of the whole system faster another valve is placed between the two pressure meters to short cut the volumes on both sides of the detector.

Sealed counters without gas flow are used mainly for radiation protection purposes and can be used for many years without gas replenishment provided they are properly conditioned during construction. This procedure consists of a repeated cycle of pumping and filling the counter with TE gas allowing the counter to steep in the gas at atmospheric pressure for several days followed by pumping down to around 0.133 mPa (10^{-6} torr). This procedure helps ensure that the level of adsorbed gas is a minimum and that the gas molecules that are adsorbed into the wall of

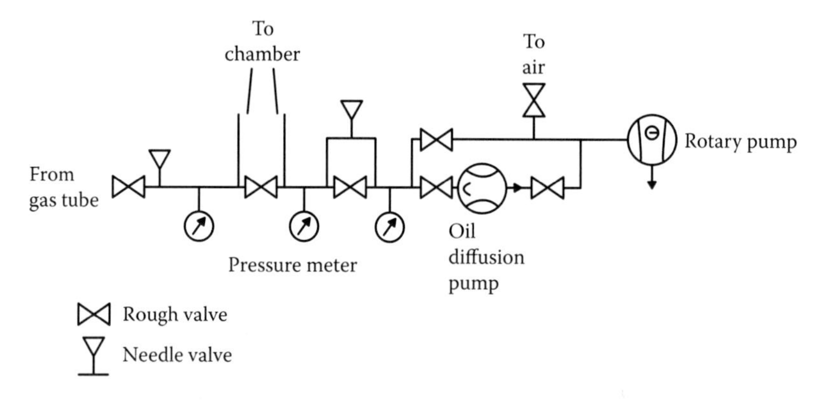

FIGURE 3.18 Schematic of a gas flow system.

the counter are those from the TE gas mixture being used rather than oxygen or water vapour. The release of gases from the detector wall can be enhanced by heating the detector to a temperature of 50°C during pumping to high vacuum. Melinder (1999) used 50°C for 12 h with improved long-term stability of the gas multiplication factor (gas gain). NASA has developed a procedure in which the detector is held at vacuum and an elevated temperature for approximately 30 days before filling. With this technique, the gas gain has been reported to be maintained for 3 years (Braby, 2015).

Chiriotti et al. (2016) have proposed an experimental method to estimate the actual size simulated by performing two independent single-event measurements, one in a gamma radiation field and one in a neutron field. In the measured dose distributions two marker points related to the electron edge and proton edge region called $h_{\mathrm{TC}}^{(e)}$ and $h_{\mathrm{TC}}^{(p)}$ are defined. Their ratio, denoted (p-edge/e-edge), was shown to be a function of the simulated spherical diameter, d, between 0.5 and 3 μm and

$$d = Ke^{0.29\left(\frac{\text{p-edge}}{\text{e-edge}}\right)} \tag{3.46}$$

for both a spherical and a cylindrical TEPC filled with either propane-based TE gas or pure propane gas. The method is one way of keeping the quality under control in measurements with a TEPC without involving calibration.

3.8 SUMMARY

In this chapter we have described a simulation principle in which a detector filled with tissue equivalent gas may replace unit-density soft tissue in the size range from about 0.1 μm and upwards. Important dimensional and dosimetric relationships between the simulated tissue volume and the simulating detector have been described along with details of the TE materials, gases and solids that are used in experimental microdosimetry. We have also referred to papers discussing how well a gas can represent real tissue as well as the experimental limitations that occur when the gas pressure is reduced to simulate volumes below 100 nm.

The two most frequent measurement methods in experimental microdosimetry, PHA of single-energy deposition events and VC analyses of multi-events, have been described in detail. In particular, the main experimental uncertainties have been listed. While in the PHA method a complete single event distribution is obtained, the VC method is limited to the determination of the dose-mean specific energy and the mean specific energy. The technique used for PHA in microdosimetry is very much the same as is used in traditional PHA gamma spectrometry and an energy calibration of the complete system is required. The many different calibration methods developed for experimental microdosimetry for energy imparted or lineal energy are described and the growing importance of intrinsic methods of calibration has been discussed. In the VC method the multi-events are usually collected as electric charge, which is converted to energy imparted via W/e. For calibration the dimensions of the gas cavity are needed as well as a calibrated electrometer. Both the PHA and VC methods use TEPCs, but in the latter technique also detectors without gas multiplication such as ionisation chambers are used. This is particularly the case when volumes in the nanometre range are simulated. Owing to the importance of TEPCs for both measurement techniques, the various designs for this class of detector that have been established and have come into common usage have been described and discussed along with a classical treatment of gas multiplication. Important particular developments such as wall-less counters and miniature TEPCs have also been discussed along with brief descriptions of a number of novel and innovative methods of radiation detection that have been adopted into experimental microdosimetry practice.

CHAPTER 4

Microdosimetry Measurements in Radiation Biology and Radiation Therapy

4.1 INTRODUCTION

The direct impact of experimental microdosimetry on the clinical practice of radiation therapy has been modest and care must be taken not to overstate its value. However, different radiation weighting factors are needed to correct for differences in biological effectiveness whenever radiation beams of different quality or linear energy transfer (LET) are used in therapy. The measurement of lineal energy dose distributions or their averages has been investigated for this purpose and found to be of considerable pragmatic value. Experimental microdosimetry is of particular interest when results from different clinics are to be compared or when changes of radiation quality are introduced. This has been found to be the case even if there is no agreed on direct relationship between measured microdosimetric quantities and a therapeutic beam's actual clinical biological effectiveness. In this chapter an attempt is made to illustrate better the connection between microdosimetric single-event distributions or their mean values and results observed in radiobiology and radiation therapy. The examples chosen reflect the interests of the authors and are by no means meant to be comprehensive or complete.

4.2 HISTORICAL NOTES ON RADIATION QUALITY

One year after Conrad Roentgen's discovery of x-rays in 1895, the first treatment of a cancer patient with x-rays took place. Although different methods had been developed for the detection of the radiation, an empirical method called the Haut erythema dose (HED) was soon used to quantify the treatments. In this method, a small area of the skin on the medical doctor or a staff member was irradiated and the exposure needed to create a certain reaction in the skin was named HED. This test was carried out in each clinic and for each of the different radiation qualities used (Lindell, 1996, 2004; Walstam, 2002).

A physical method to measure exposure was soon developed and a unit was defined in Germany that was called the R-unit. The method was based on the number of ionisations created by the radiation inside a defined volume of air at a specified temperature and pressure. In the 1920s several national investigations of the dose given to patients and based on HED were compared with the dose based on the new physical quantity. Differences of a factor of 4 were reported but more typically the values were within ±20% (Sievert, 1926). In many countries, authorities began to make periodic checks of the output of the radiotherapy x-ray units to improve therapeutic outcomes and consistency in radiation response. Today it is known that a difference in absorbed dose of 5% to 10% can be clinically detected and an accuracy of ±3% in absorbed dose is strived for.

X-rays generated by potentials up to about 300 kV dominated external radiation therapy for a long period. In the 1950s ^{60}Co γ as well as ^{137}Cs γ sources of sufficient strength for cancer treatments became available for radiotherapy. Also, about this time, dedicated accelerators began to be produced for cancer treatment. It was observed fairly early on that to achieve the same treatment results with high-energy photon beams as with x-rays the absorbed dose to the target had to be increased by about 25%, which demonstrated that the dose quantities in use were not sufficient on their own for predicting the outcome of a treatment. Weighting factors had to be used. Nowadays protons, neutrons and ions are available for treatments and other values of the dose-weighting factor are needed (IAEA, 2008).

4.3 RELATIVE BIOLOGICAL EFFECTIVENESS

4.3.1 Relative Biological Effectiveness in Single-Dose Cell Survival Experiments and in Dose-Fractionated Radiation Therapy

When cells are irradiated, a fraction of them will not survive and the surviving fraction will diminish with increasing absorbed dose. A common relation describing the fraction of surviving cells, $S(D)$, after the absorbed dose D, relative to the survival at zero dose determined by a measurement of the cell line plating efficiency, $S(0)$, is

$$\frac{S(D)}{S(0)} = e^{-\left(\alpha D + \beta D^2\right)} \tag{4.1}$$

Here α and β are the linear and quadratic survival parameters and at low doses α is the initial slope of the survival curve. The α term is supposed to represent damage induced by single tracks of ionising particles (Barendsen, 1990). The frequency of cellular effects is given by

$$F(D) = \alpha D + \beta D^2 \tag{4.2}$$

The fraction of cells surviving a specific absorbed dose is also dependent on the radiation quality as illustrated in Figure 4.1. To quantify the change needed in absorbed dose when going from one radiation quality to another, the relative biological effectiveness (RBE) was introduced. It is the ratio of the two absorbed dose values at which the cell survival fraction is the same. It can be calculated for any desired level of survival or any kind of biological endpoint of interest. If, in the figure, the RBE at 10% cell survival is compared with the RBE at 1% survival, we observe that the RBE values are different, 5.7 as compared to 4.7 in the latter. It is thus important in studies of RBE to specify what biological system and which biological endpoints are being considered.

Another common characteristic of RBE is its dependence on LET. Such a relation is shown in Figure 4.2. Here RBE starts to increase at about 10 keV μm^{-1} and passes through a maximum around 100 keV μm^{-1} and then decreases. The decrease in RBE after the peak is interpreted in the following way. Once sufficient energy has been deposited to initiate the maximal effect any larger amount of energy will consequently result in a reduction of the effect per energy unit. This is sometimes called the saturation effect or, more colourfully, as 'over-kill'. Again we see that the RBE is a dose-dependent quantity and the peak is becoming less pronounced with increasing absorbed dose.

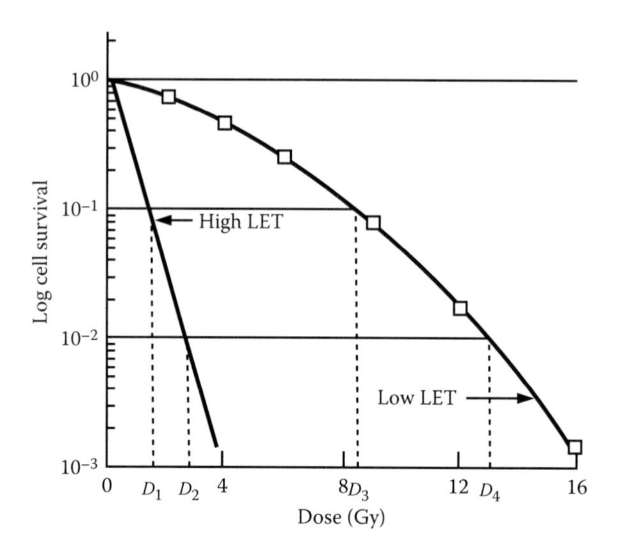

FIGURE 4.1 Relative cell survival as function of the absorbed dose to the cells. The line with open squares is cell survival after irradiation with a low-LET beam such as ^{60}Co γ-rays. The line without markers is cell survival after irradiation with a high-LET beam. Relative biological effectiveness (RBE) is defined as the ratio of the absorbed dose values for which, in this case, the cell survival is the same. The dose values for 10% and 1% cell survival are marked in the figure as D_1 and D_3 respectively D_2 and D_4. RBE for the two levels of survival becomes $(D_3/D_1) = (8.5/1.5) = 5.7$ and $(D_4/D_2) = (13/2.75) = 4.7$. The figure illustrates that RBE is dependent on the endpoint chosen. Here RBE for 10% survival is larger (5.7) than that for 1% survival (4.7). Modified from ICRP (International Commision on Radiologocal Protection). The 2007 recommendations of the International Commission on Radiological Protection. Publication 103. *Annals of the ICRP* 37 (2–4) (2007).

Since the 1970s, much clinical experience has been gained from treatments with ionising radiation with different fractionation schedules. These are summarised by a linear quadratic (LQ) relation (IAEA, 2008) and

$$BE = \alpha n d + \beta n d^2 \tag{4.3}$$

or

$$BE = \alpha D\left(1 + \frac{d}{\frac{\alpha}{\beta}}\right) \tag{4.4}$$

Here the biological effect (*BE*) is related to the dose fraction, *d*, of the absorbed dose, *D*, delivered in *n* equally large fractions, and α and β are the

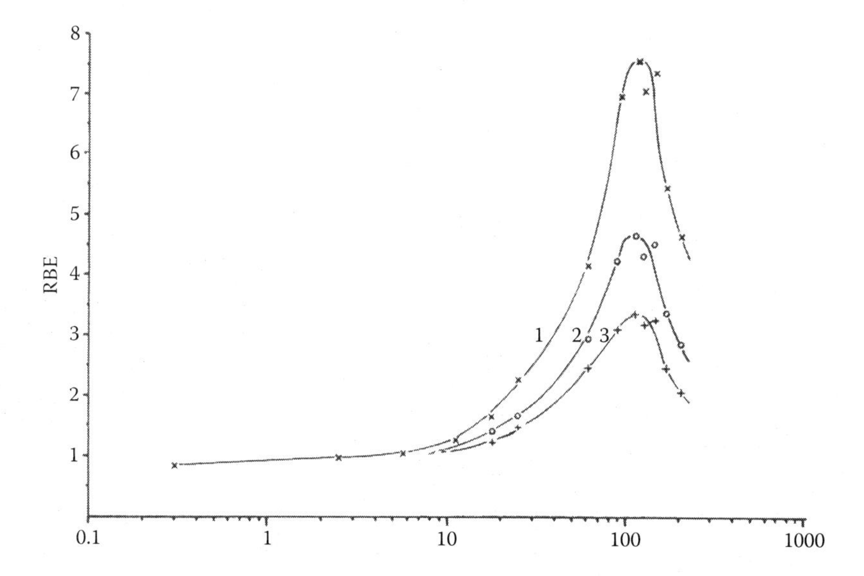

FIGURE 4.2 Variation of RBE with LET for survival of mammalian cells of human origin. Curves 1, 2 and 3 refer to cell survival levels of 0.8, 0.1 and 0.01. (Reproduced from Barendsen G. W. Responses of cultured cells, tumours and normal tissues to radiations of different linear energy transfer. *Current Topics in Radiation Research Quarterly* 4:293–356, 1968.)

linear and quadratic parameters of the survival curve for the cell line being treated. The preceding relation is expected to be valid for dose fractions in the range 2 to 10 Gy (Brenner, 2008). As the first term is not much affected by changes in fractionation, it is assumed to be due to single-energy deposition events. Independent of which radiation quality is chosen for a treatment, the same endpoint, *BE* is the goal. The LQ relation may be used for predicting the RBE value or weighting factor (IAEA, 2008) applicable to another radiation therapy beam, *b*, if α and β are known for both radiation qualities. If the reference radiation is ^{60}Co γ distributed as d_γ Gy until the total dose D_γ is reached, and if the alternative radiation quality is distributed as d_b Gy until the total dose is D_b Gy, the weighting factor called W_b is

$$W_b = \frac{D_\gamma}{D_b} = \frac{\alpha_b\left(1 + \frac{d_b}{\left(\frac{\alpha}{\beta}\right)_b}\right)}{\alpha_\gamma\left(1 + \frac{d_\gamma}{\left(\frac{\alpha}{\beta}\right)_\gamma}\right)} \tag{4.5}$$

TABLE 4.1 Clinically Observed Radiation Weighting Factors, W_b

Radiation Quality	W_b
x-ray/^{60}Co γ	1.25
p (175 MeV)/^{60}Co γ	1.1
(in SOBP at 5 cm)	IAEA (2008)
^{12}C (300 MeV u^{-1})/^{60}Co γ	2.4
(centre of SOBP)	Matsufuji et al. (2007)
^{12}C (300 MeV u^{-1})/^{60}Co γ	3.0
(distal end of SOBP)	Matsufuji et al. (2007)
Neutrons/^{60}Co γ	3.2
	Batterman et al. (1981)

Note: The therapy beams were conventional x-rays, ^{60}Co γ, a neutron therapy beam, a proton beam (290 MeV) and a carbon ion beam (290 MeV u^{-1}). SOBP, Spread Out Bragg Peak.

Weighting factors derived from clinical observations, W_b, for several radiation qualities and early reacting tissues and tumours are given in Table 4.1.

The two equations, Equation 4.2 for single-dose cell survival experiments in vitro and Equation 4.3 for dose-fractionated radiation therapy, are similar. Although α and β appear in both they are not necessarily the same. However, the α coefficient in both expressions is usually expected to be independent of the absorbed dose but dependent on radiation quality, meaning that it is reflecting damage caused by single events. A relation or connection between the α coefficient and single events therefore seems likely. The β coefficient is sometimes considered to be independent of radiation quality.

4.3.2 Radiobiological Experiments in Support of Sensitive Targets in the Nanometre Range

A number of radiobiological experiments support the existence of important sensitive targets in the nanometre range:

1. Mammalian-cell survival curves were analysed after irradiation with deuterons and alpha particles. From these results, the size of a sensitive target was estimated to be 7 to 10 nm (Barendsen, 1967, 1990). This value was later revised to be 20 nm.

2. In an elegant experiment known as the molecular ion experiment at Columbia University in the late 1970s the objective was to investigate interaction distances for radiation-induced damage. In

this experiment, correlated pairs of ions produced by the fragmentation of a molecular beam of deuterons traversed cells in tissue culture with varying separation. References linked to this experiment are found in Kellerer et al. (1980). Analyses by Zaider and Brenner (1984) showed that sub-lesions interact primarily over distances of less than about 20 nm but such interactions may occur up to a few micrometres but with lower probability. This experiment was performed for V79 Chinese hamster cells in late S phase.

3. Ultra-soft x-rays (x-rays of energies from 0.3 to about 5 keV) with electron ranges from about 7 nm to about 600 nm were used for inactivation of V79 cells (Goodhead and Nikjoo, 1989). In spite of the very short electron range (7 nm) in the carbon K x-ray beam, a high efficiency for inactivation was observed resulting in an RBE of 3.4 relative to hard x-rays. However, different cell lines expressed different RBEs.

4. Brenner and Ward (1992) concluded that clusters of two to five ionisations within a volume of 1 to 4 nm best fitted RBEs for induction of a double-strand break (DSB) at different LETs.

5. A linear relation between the α value yield of dicentric chromosomes in human lymphocytes and $\bar{L}_{100,D}$ was reported by Bartels and Harder (1990). The results were obtained after irradiations in photon, electron and a few neutron beams with LET below 70 keV μm^{-1} (Edwards et al., 1990). The values of $\bar{L}_{100,D}$ were calculated with a Monte Carlo track-structure code. The resulting number of ionisations in a cylinder of infinite length and with a diameter of either 10 nm or 20 nm was converted to energy imparted through multiplication with 40 eV per ionisation and the mean value $\bar{L}_{100,D}$ was arrived at as described in Chapter 2, Section 2.3.2. An alternative to $\bar{L}_{100,D}$ was suggested to be $\bar{y}_D$ and a volume with 25 nm diameter.

6. A model for predicting the yields of DSBs from a measured ionisation cluster distribution in a volume 4.3 nm in diameter and 6 nm long was presented by Garty et al. (2010). In the model, the probability to form DSBs was taken from experiments with plasmids. The predicted yield of DSBs exhibited the same dependence of LET as published results but the values of the DSBs was a factor of 2 to 3 above reported values. The lack of repair processes was a reason given for this difference.

These examples illustrate that it is likely that there exist critical targets contained within nanometre volumes. This does not exclude the possibility that larger volumes may also be of importance.

As we have seen, measurement techniques for microscopic volumes are well established, while techniques suitable for nanometres volumes are few. Track-structure calculations, on the other hand, can be used to evaluate energy deposition on the nanometre scale, avoiding experimental problems. Track-structure codes, however, have their own difficulties and confidence is gained by having calculations and measurements agree within the limits of their associated uncertainties.

4.3.3 Dose Range and Volumes in Which Single Events Dominate

The dose range in which single events dominate for a particular radiation quality can be estimated using Poisson statistics (Chapter 2), provided $\bar{z}_s$ is known. The results for 100 keV electrons are shown in Table 3.3 for two different cylindrical volumes with diameters and heights equal to 10 nm and 100 nm, respectively. Values of $\bar{z}_s$ were calculated from the graph in Figure 2.11, presenting $\bar{y}$ for such cylinders as a function of electron energy. The last line of Table 4.2 $[1 - p(0) - p(1)]$ shows the probability for more than one event. This probability is small at 10 nm for both 2 and 60 Gy. At 100 nm and 2 Gy single events are still more frequent than multiple events, while at 60 Gy the probability for multiple events is about four times larger than the probability for single events, $p(1)$. For radiation beams of higher LET (larger $\bar{z}_s$) it is obvious that the volumes in which single events dominate will be larger than for 100-keV electrons at the same absorbed dose. We may conclude that in all therapy beams and

TABLE 4.2 Dose Ranges in Which Single Events are Dominant for 100 keV Electrons. Mean Specific Energy, $\bar{z}_s$, Calculated from the Mean Lineal Energy, $\bar{y}$, in Figure 2.11 and for Cylinders of Equally Large Diameter and Height

	Cylinder Diameter			
	10 nm		100 nm	
$\bar{z}_s$	21.9 Gy		9250 Gy	
D	2 Gy	60 Gy	2 Gy	60 Gy
$p(0)$	1.00	0.994	0.913	0.065
$p(1)$	2.2×10^{-4}	0.006	0.083	0.177
$[1 - p(0) - p(1)]$	2×10^{-7}	3×10^{-5}	00.004	00.758

Note: For 100 keV electrons the values are $\bar{y} = 6.8$ keV μm^{-1} at 10 nm and 1.61 keV μm^{-1} at 100 nm.

volumes with diameters of less than about 100 nm single events alone can be expected to cause the cellular damage at dose values less than 2 Gy. At 60 Gy single events will still dominate at 10 nm.

An alternative method to judge the importance of the energy imparted by single events as opposed to multi-events is to calculate the integral proximity function, $T(x)$ (ICRU, 1983). The integral proximity function (Chapter 2, Section 2.4) defines the mean energy imparted, $\bar{\varepsilon}$, to a spherical target of radius x, centred on an arbitrary transfer point on an arbitrary particle track. Figure 4.3 shows this integral for ^{60}Co γ and 15-MeV neutrons as a function of target radius.

In Figure 4.3 we find that on average a single event from a beam of 15-MeV neutrons deposits roughly 10 times more energy within a sphere with a radius of 10 nm than an electron created by ^{60}Co γ radiation.

If a tissue with density ϱ kg m^{-3} has been given the absorbed dose D Gy, the mean energy imparted is

$$\bar{\varepsilon} = Dm = 4D\pi x^3 \varrho / 3 \tag{4.6}$$

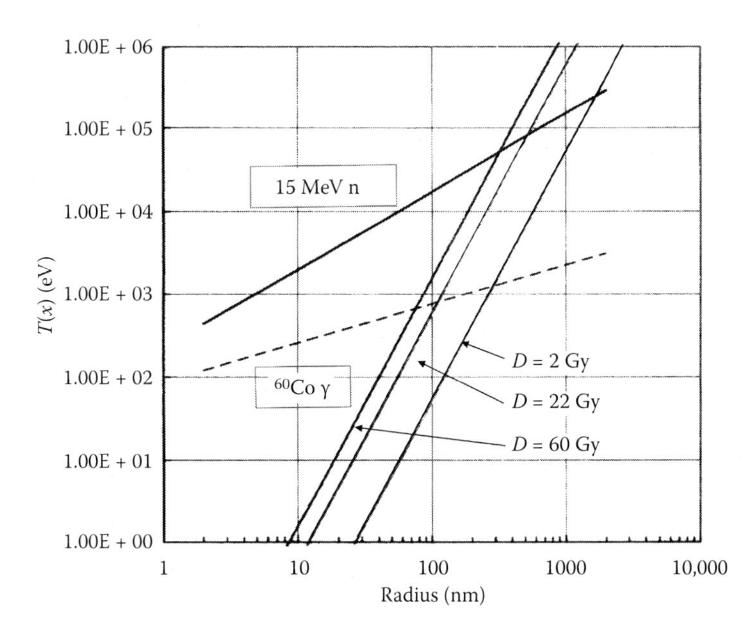

FIGURE 4.3 The integral proximity function, $T(x)$, for various radii, x, and 15-MeV neutrons and ^{60}Co γ. Also shown is the mean energy imparted within various radii at 2 Gy, 22 Gy and 60 Gy. (From Lindborg L. and Grindborg J. E., *Radiat. Prot. Dosim.* 70(1/4), 541–546 (1997).)

For a given dose, D, the mean energy deposited, $\bar{\varepsilon}$, increases rapidly with the radius, x. In Figure 4.3 $\bar{\varepsilon}$ (in eV) is also plotted as function of x for three common dose values used in radiation therapy, 2 Gy, 22 Gy and 60 Gy. $T(x)$ will be larger than $\bar{\varepsilon}$ calculated from D for small values of x, because $T(x)$ is defined for an individual charged particle and its energy transfer points, while D is an averaged quantity for an entire exposure and, for volumes with small radii, will be increasingly affected by so-called zero-events in which no energy deposition takes place. When in the figure $T(x) > \bar{\varepsilon}$, the mean energy imparted is expected to be deposited by a single track or event. Thus for 2 Gy single events dominate by far up to a radius of 300 nm for ^{60}Co γ and up to at least 1000 nm in the neutron beam. At 60 Gy and ^{60}Co γ single events are expected to dominate up to about 70 nm. Even though Figure 4.3 shows results for spherical volumes characterised by their radii and in Table 4.2 the volumes are cylinders characterised by their diameters, the conclusions drawn from both sets of data support each other very well.

4.3.4 Correlations between Oncogenic Transformation and Measures of Radiation Quality for Different Site Sizes

Oncogenic transformation in the C3H 10T1/2 cell system was studied in mono-energetic neutron beams from 200 keV to 14 MeV as well as in beams of charged particles with LET values from 10 keV μm^{-1} to 150 keV μm^{-1} and a 250 kV x-ray beam. The induced transformation frequencies per surviving cell as a function of dose, D, were observed for the different beams, i, and all data were fitted using the maximum likelihood criterion to a linear quadratic model:

$$Y_i(D) = \alpha_i D + \beta D^2 \tag{4.7}$$

Here α_i and β are $n + 1$ free parameters for the n different radiations. Little improvement was observed if also β was allowed to vary. The α_i was taken to represent the initial slopes of the data.

The radiation beams were characterised by different mean values for quantities representing radiation quality; of particular interest were $L_{T,\infty}$, $L_{D,\infty}$, $L_{D,100}$, $\bar{y}_D$. The last quantity was evaluated both at 1 μm and 25 nm. The RBE_{max} was found to be best predicted by L_D and $\bar{y}_D$ at any of the simulated volume sizes (Hall et al., 1990).

4.3.5 Correlations between RBE for Low Energy Photons and Electrons and Measures of Radiation Quality for nm-Site Sizes

A review of RBE_{max} for several biological endpoints after irradiation with low-energy electrons and photons was reported by Nikjoo and Lindborg (2010). The study included chromosome aberrations, cell survival and experimentally determined as well as calculated DSBs observed for a $^{60}Co\ \gamma$ beam as well as electron and photon energies between 0.3 and 10 keV. RBE_{max} (calculated as the ratio α_x/α_γ) varied from 1.1 to 7.7 with an average RBE_{max} of 3 ± 2 (1 standard deviation). The ratios $L_{100,D,x}/L_{100,D,\gamma}$ and $\bar{y}_{D,x}/\bar{y}_{D,\gamma}$ for the different beams were used as predictors of these RBE values. However, when the α ratios were divided by these predicted RBEs (for a cylinder with 10 nm diameter and height) for the particular radiations involved, the resulting quotients tended towards unity but the standard deviation remained large (Table 4.3). As several different endpoints were included, this observation may not be surprising. Also, the uncertainty in biological experiments is usually quite large. If the same calculations are carried out for the DSBs alone, the agreement between the observed and the predicted RBEs is improved (Table 4.3).

4.4 SINGLE-EVENT DOSE DISTRIBUTIONS

Single event distributions for volumes in the micrometre range have been reported for many beams used in radiation therapy and radiation biology. As the focus of these studies was usually on energy imparted, lineal energy dose distributions have been of particular interest.

The simplest assumption concerning a relation between a single-event dose distribution and RBE is that if the same distribution is measured in two different beams or at different positions in the same beam, the expected RBEs will be the same. In practice, it is very unusual to find two identical distributions even at two different depths in the same beam, and consequently a more quantitative judgement is required in assessing how

TABLE 4.3 Mean RBE Values of *RBE* in the Literature for Cell Survival, Chromosome Aberrations and DSBs after Irradiations with Low-Energy X-Rays (0.3 keV to 8 keV) and $^{60}Co\ \gamma$

Biological Effect	$\overline{\mathbf{RBE}_{max}}$	$\frac{\alpha_x/\alpha_\gamma}{L_{100,D,x}/L_{100,D,\gamma}}$	$\frac{\alpha_x/\alpha_\gamma}{\bar{y}_{D,x}/\bar{y}_{D,\gamma}}$
Cell survival, chromosome aberration and DSBs	3 ± 2	1.2 ± 1	1.7 ± 1
DSBs only	2.0 ± 0.5	0.8 ± 0.2	1.1 ± 0.5

Source: Nikjoo H. and Lindborg L., *Phys. Med. Biol.* 55, R65–R109 (2010).

equivalent the beams will be in biological effect. It is important to realise that a comparison of mean values alone, such as values of $\bar{y}_D$, may be misleading, as the biological efficiency in different y-intervals may be different. Therefore, it is also important to involve biological and/or clinical experiences when interpreting microdosimetric information. Employing more sophisticated experimental techniques can also be useful to understand radiation quality and its relationship to RBE such as using time-of-flight measurements to determine the dose fractions from different charged particles (Endo et al., 2010). Figures 4.4 and 4.5 show that quite large differences in lineal energy distributions can be found between different radiation qualities for volumes in the micrometre range (Eivazi, 1990; Lindborg and Nikjoo, 2011). This is an argument in favour of why such distributions are useful for characterising radiation beams used in clinical or radiobiological investigations.

4.4.1 Qualitative Judgements of RBE Using Single Event Dose Distributions

When high-energy electron beams were introduced for radiation therapy around 1970, a possible increase in RBE at the end of the depth-dose curve was considered. The reason was the large number of electrons that will stop at roughly the same depth. As LET will increase with decreasing energy, an increase in RBE could not be excluded. Figure 4.6 shows that indeed a shift in the single-event dose distribution towards larger events is observed with increasing depth in a beam of 15-MeV electrons. However, this increase is small compared to the difference observed between the single-event spectra

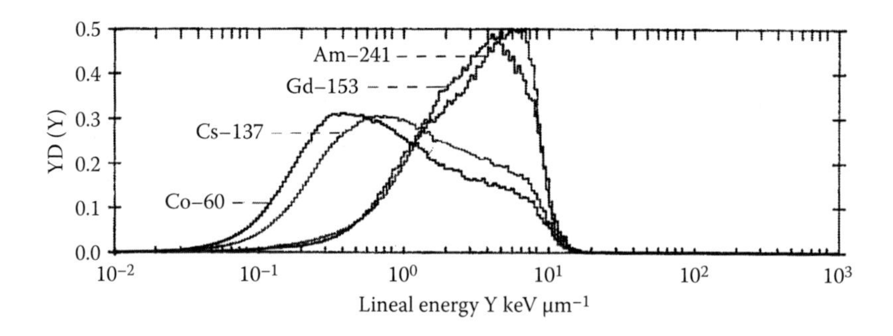

FIGURE 4.4 Single-event lineal energy dose distributions for different photon energies measured with a Far West Technology 1.27 cm Rossi counter simulating a 2-μm tissue sphere.

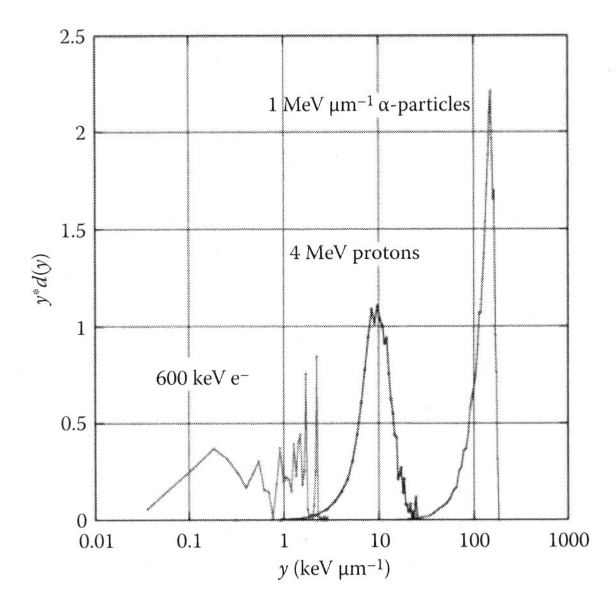

FIGURE 4.5 Lineal energy dose distributions for three different radiation qualities: 600 keV electrons, 4 MeV protons and 1 MeV μm^{-1} alpha particles. Calculations were made for a unit density sphere with 1 μm diameter. (From Lindborg L. and Nikjoo H., *Radiat. Prot. Dosim.* 143, 402–408 (2011).)

similarly measured for electron beams at three different energies including 15 MeV and for ^{60}Co γ rays (Figure 4.7) (Lindborg, 1976).

If there is no RBE difference between the electron beam and the gamma beam at small depths, an increase of RBE with depth in the electron beam is less likely to be observed. Today the same RBE is assumed along the depth-dose curve in treatments with high-energy electron beams.

A detailed study of single-event dose distributions measured in a pion beam and at several different depths was reported by Schuhmacher et al. (1979) (Figure 4.8). Of particular interest in their study was the investigation of a possible change of the radiation quality when the Bragg peak was broadened by an increase of the momentum spread. The measured lineal energy dose distributions were divided into three y-intervals: $y < 5$ keV μm^{-1}, $5 < y < 150$ keV μm^{-1} and finally $y > 150$ keV μm^{-1} and the dose contributions in those intervals were determined for the different depths (the dose contribution in a particular y-interval is proportional to the area below the curve in that y-interval). The measurement series were repeated for two different settings of the spread of the dose peak and the results are shown in Figure 4.9a and b. For the broader beam the dose

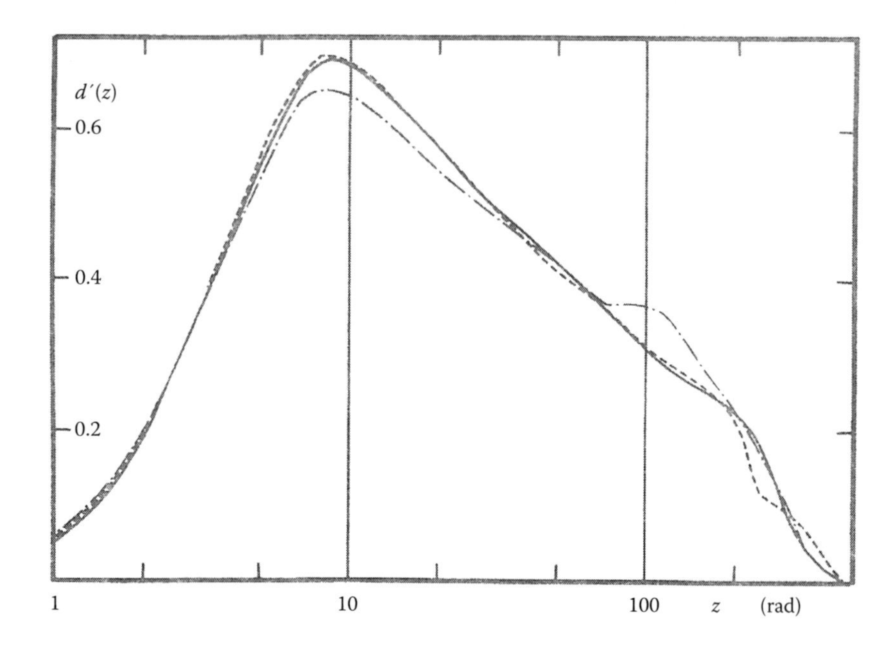

FIGURE 4.6 Specific energy dose distributions for a beam of 15-MeV electron energy measured at three different depths: – – –, 0.5 g cm^{-2}; —, 2.4 g cm^{-2} and — · —, 5.7 g cm^{-2} (Lindborg, 1976). The measurements were made with a spherical TEPC simulating a volume of 0.95 μm diameter. In this representation ($d'(z) = zd(z)$ vs. $d \ln z$); equal areas under the curves correspond to equal specific energy contributions.

fractions due to intermediate and high lineal energies is smaller. A higher biological effectiveness was therefore expected in the narrow beam and this was found to be in agreement with radiobiological experiments.

4.4.2 Single-Event Dose Distributions and Boron Neutron Capture Therapy

A way of increasing the absorbed dose to tumour cells is to enrich them with boron (B) and then irradiate the tumour with thermal or epithermal neutrons. Thermal neutrons have a large cross section for capture reactions with boron after which an alpha particle is emitted along with a Li recoil nucleus. The higher LET of these particles is assumed to make them more efficient in killing the tumour cells and hence have a larger RBE compared to the low-LET radiation used more routinely for therapy. Attempts have been made to determine the dose fractions from the different particles using microdosimetric techniques.

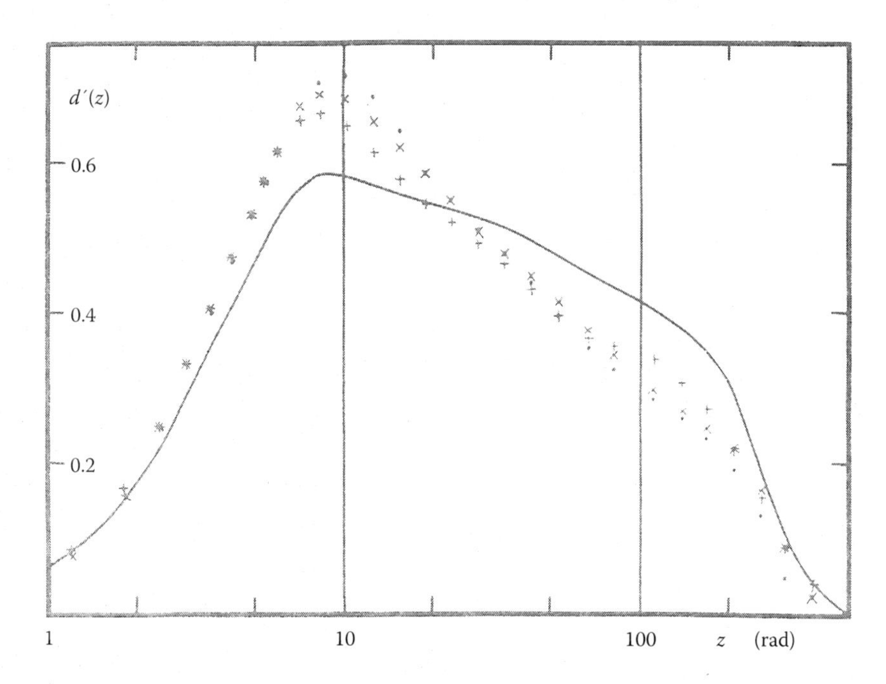

FIGURE 4.7 Single-event specific energy dose distributions for electron beams. ···· 39 MeV measured at 3.4 g cm^{-2}, xxxx 15 MeV measured at 2.4 g cm^{-2} (solid line in Figure 4.8) and ++++ 42 MV x-rays measured at 5.1 g cm^{-2} and solid line, ^{60}Co γ-rays measured at 1.2 g cm^{-2} (Lindborg, 1976).

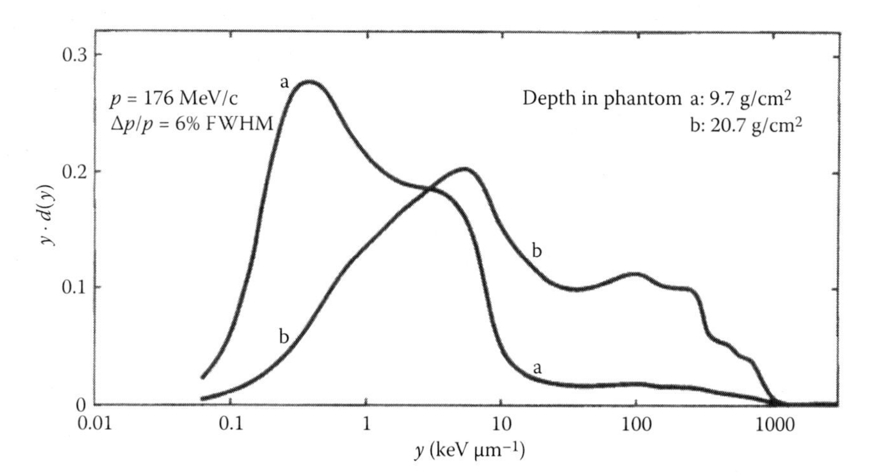

FIGURE 4.8 Dose lineal energy distribution measured in a pion beam momentum 176 MeV/c and momentum spread 6% full width at half maximum (a) in the plateau region and (b) peak region. The simulated volume had a diameter of 2 μm. (From Schuhmacher H. et al., *Radiat. Environm. Biophys.* 16, 239–244 (1979).)

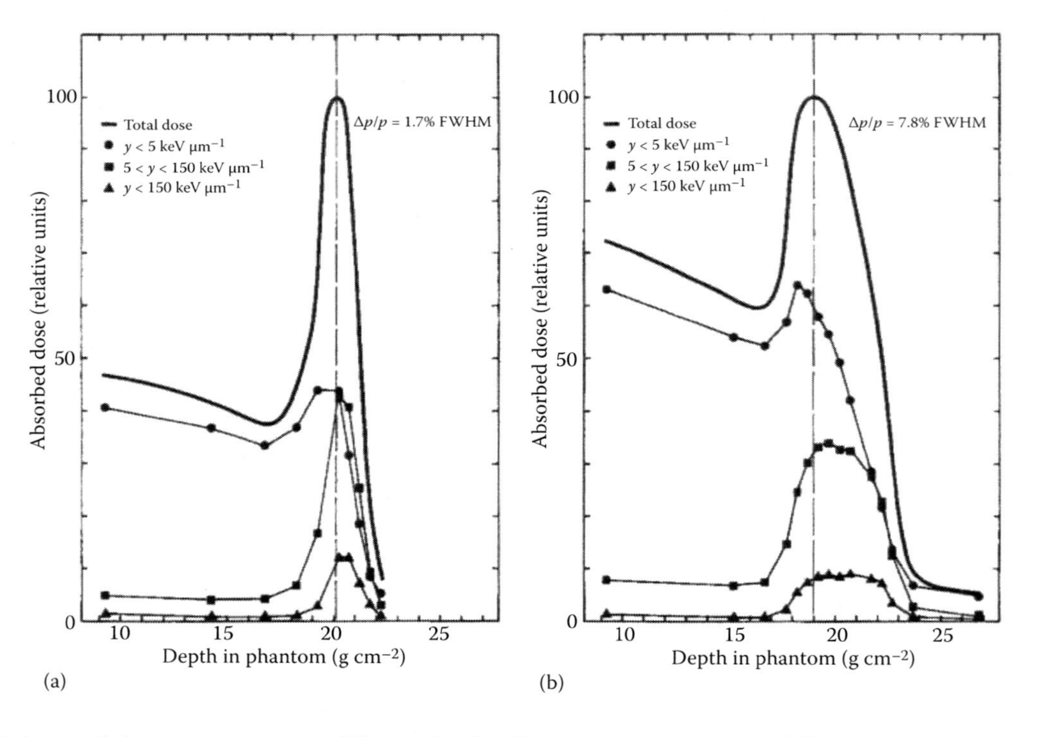

FIGURE 4.9 Total dose and dose components at different depths. The dose components in different intervals of y were calculated from the dose lineal energy distributions and the beam in Figure 4.10. In (a) momentum spread was 1.7% (full width at half maximum [FWHM]); in (b) the momentum spread was 7.8%. When compared the dose fractions due to high and intermediate high lineal energies are smaller when the momentum spread is larger. (From Schuhmacher H. et al., *Radiat. Environm. Biophys.* 16, 239–244 (1979).)

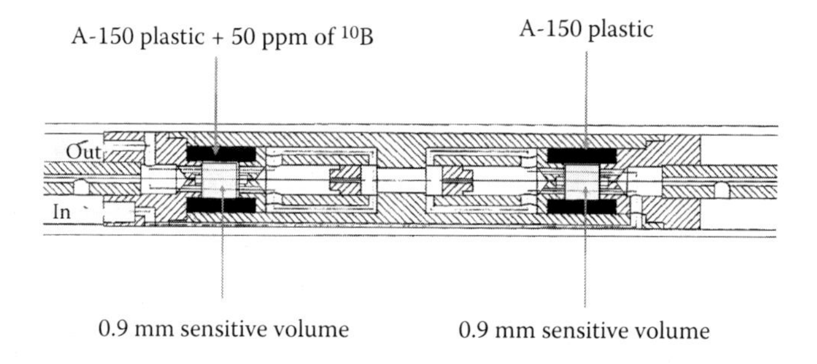

FIGURE 4.10 **(See colour insert.)** The twin chamber for measurements in a neutron beam for BNCT. The TEPC is a square-cylinder with a diameter and height of 0.9 mm. The A-150 wall in one of the chambers has been doped with 50 ppm of ^{10}B. The boron atoms may capture thermal neutrons and emit alpha particles along with a Li ion recoil. In BNCT the tumour is enriched with boron and the lineal energy distribution measured with the boron-doped TEPC will be similar to that experienced by the tumour cells. (From Moro D. et al., *Appl. Radiat. Isotop.* 67, S171–S174 (2009), Figure 1.)

A twin chamber constructed for boron neutron capture therapy (BNCT) measurements is shown in Figure 4.10. This counter was designed with one detector as a regular tissue-equivalent proportional counter (TEPC) while the other had a TEPC wall enriched with 50 ppm of ^{10}B. The standard TEPC gives a spectrum of gamma and neutron dose components, while the detector with the boron-loaded wall also includes the dose component from ^{10}B fragments (Moro et al., 2009). Figure 4.11 shows the total dose distribution as well as the different dose components when the detector simulates a cylinder with 1 μm diameter. The measurements were compared to the results of Monte Carlo calculations (MCNPX 2.6b code). The measured neutron doses were the same, while the experimental gamma dose was significantly larger than the calculated value. The reason for this was suggested to be insufficiently precise modelling of the detector. The dose contribution from the boron fragments was smaller than the calculated value and thought to be due to partial absorption of the boron fragments inside the metallic boron powder used to enrich the A-150.

4.5 RESPONSE FUNCTIONS DERIVED FROM SINGLE-EVENT DOSE DISTRIBUTIONS

4.5.1 Response Function Derivation

During the 1980s a number of reports appeared describing large variations (up to 50%) in RBE between different neutron therapy beams as well

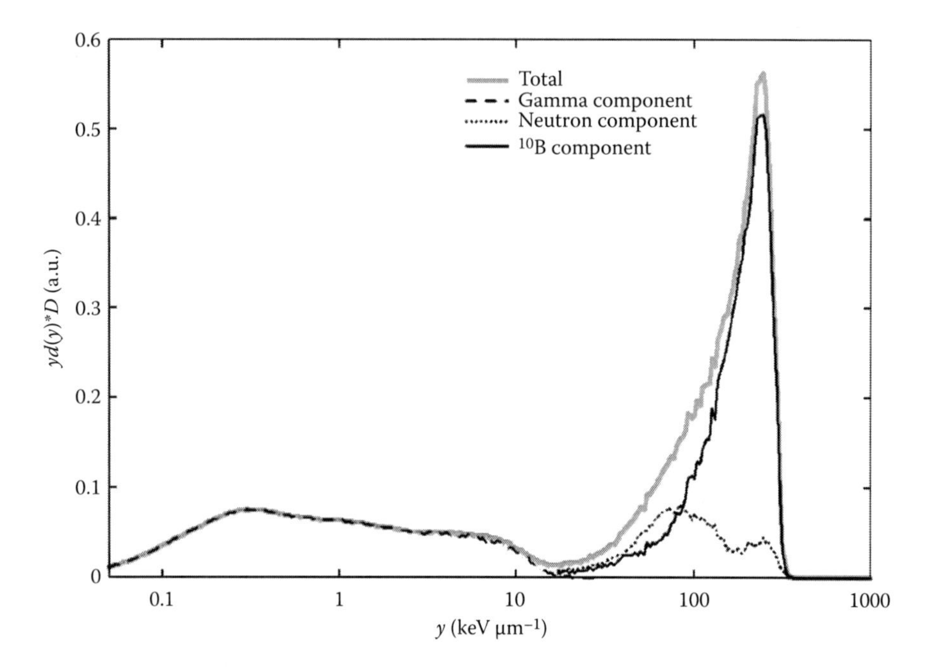

FIGURE 4.11 Dose components obtained with the twin TEPC detectors. The 'total' lineal energy dose distribution obtained from the boron-doped TEPC is marked in grey. The gamma and neutron dose components measured with the non-doped TEPC are shown as dashed and dotted lines. The difference between the two detectors reveals the component of the total dose due to boron capture (solid black line). For details see Moro et al. (2009), Figure 6.

as differences with depth. To improve understanding of radiation quality variations in neutron therapy beams, a microdosimetric intercomparison of the neutron therapy facilities was organised by the Heavy Particle Therapy Group of the European Organisation on Research and Treatment of Cancer (EORTC) based on the earlier work of Menzel (1981) and conducted mainly by Pihet (1989). A review has been presented by Menzel et al. (1990). Single-event dose distributions (measured at 2 µm) from 14 different neutron beams together with RBE values observed in those beams were analysed. One facility was used as the reference beam. It was then assumed that there exists a response function $r(y)$ such that a biological weighting factor R is given by

$$R = \int r(y)d(y)dy \tag{4.8}$$

In the reference beam R equals 1 and R in other beams becomes an estimate of the RBE. The calculation was made for nine of the neutron beams and for two different biological endpoints. One biological endpoint was regeneration of intestinal crypt cells in mice after 8 Gy (Pihet et al., 1990). These cells react early on following irradiation. The calculation starts with a guess function for *r*(*y*) followed by successive iterations. As a statistical test, the resulting *R* values were compared with the experimentally observed RBE values that were used as input and the ratios obtained were found to deviate from unity by less than 3% (1 relative standard deviation). These results show that the needed accuracy in *R* can be achieved using this method. Later, Loncol et al. (1994) extended the investigation to proton beams and a ^{60}Co γ beam. The same biological endpoint as in the study by Pihet et al. was used. The *r*(*y*) function could then be extended down from about 10 keV μm^{-1} to 0.01 keV μm^{-1}. In this range *r*(*y*) was found to be almost constant at a value of 1. The shape of the *r*(*y*) peak derived from the neutron beam investigations remained almost unaffected by the proton and gamma beam data. In Figure 4.12 (Wambersie et al., 2002), this extended response function, *r*(*y*), for regeneration of intestinal crypt cells in mice is shown. The shape of the *r*(*y*)

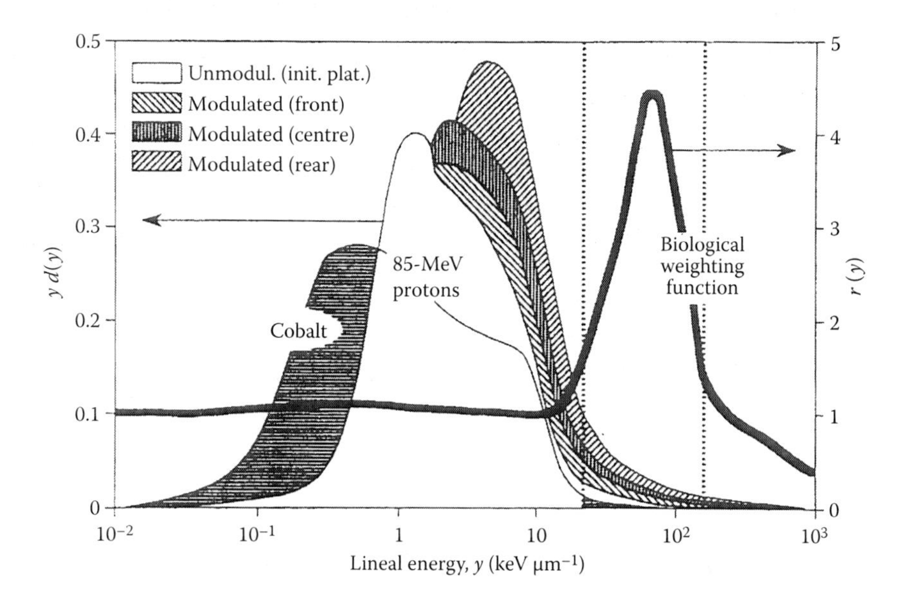

FIGURE 4.12 The biological weighting function *r*(*y*) for early effects in mice based on investigations in a ^{60}Co γ beam, proton and neutron beams (Loncol et al., 1994). Lineal energy dose distributions measured for ^{60}Co γ and protons of 85 MeV without energy modulation (white), and three points along the SOBP. (From Gueulette J. et al., *Radiat. Res.* 145, 70–74 (1996).)

function is similar to those often observed when RBE is presented as a function of LET (Figure 4.2). This is not surprising, as L and mean lineal energy are numerically close for a simulated volume of 2 μm. However, contrary to LET, the y-distributions are measurable and the method is thus of practical importance. The work of Pihet et al. (1990) showed that with measured lineal energy single-event dose distributions at 2 μm and RBE investigations, it was possible to define a response function $r(y)$, which when applied to the single-event dose distributions provides a radiation quality weighting with the needed accuracy for specific biological endpoints. A few applications of this method are described in Sections 4.5.1, 4.5.2 and 4.5.3. A similar investigation for a beam of 50 MV x-rays was published by Tilikidis et al. (1996).

Brenner and Zaider have objected to this approach of relating lineal energy and RBE at high dose values (Brenner and Zaider, 1998). They argued that for a site with a diameter of 2 μm and for low-LET radiation at an absorbed dose of 8 Gy, the situation was no longer one where single events applied (Figure 4.3). The response by Pihet and Menzel (1999) placed the emphasis on the empirical nature of $r(y)$ functions, which are therefore valid within the experimental data range from which they were derived and care must be used when applying these functions outside this input data range.

4.5.2 The Response Function $r(y)$ Applied to a Proton Therapy Beam of 62 MeV

Single-event dose distributions were determined in a proton therapy beam of 62 MeV. The measurements were made with an ultra-miniature TEPC developed by Colautti and co-workers (Cesari et al., 2000; De Nardo et al., 2004). The counter's cylindrical volume had a diameter of 0.9 mm and the same height (similar to the detector in Figure 4.10). The small physical dimensions make it valuable for investigations in the Bragg peak region of the depth-dose curve. The TEPC simulated a diameter of 1 μm. Several depths in a Lucite phantom were investigated. In Figure 4.13 the measurement positions as well as the single-event dose distributions along the depth-dose curve are demonstrated. The $r(y)$ function for early effects in mice at 8 Gy (Loncol et al., 1994) were then applied to the single-event dose distributions to determine the response called RBE_{μ} equivalent to R from Equation 4.6 at the different depths. This response is shown in Figure 4.14. It increases from almost 1.0 at the entrance of the phantom to about 2.3 close to the end of the spread-out Bragg peak (SOBP). The

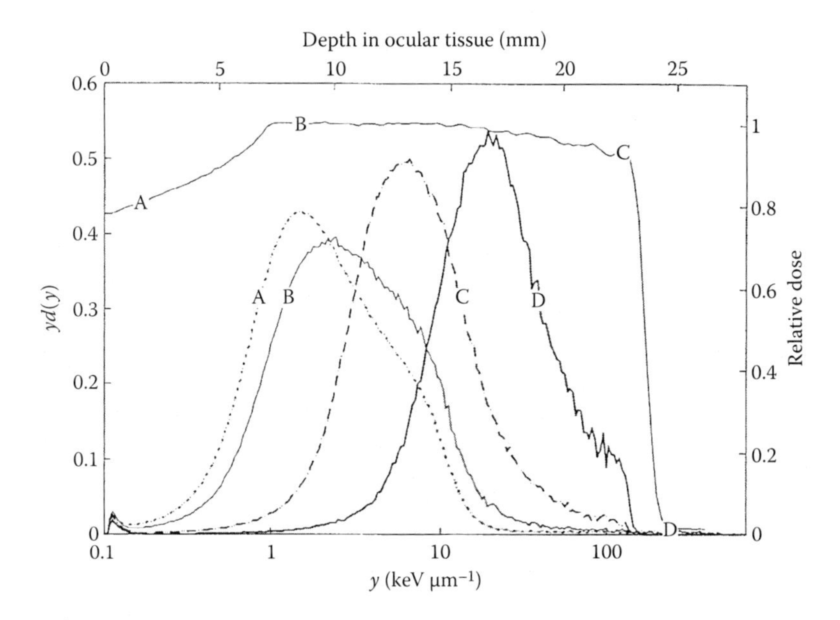

FIGURE 4.13 Lineal energy dose distributions measured in a 62-MeV proton therapy beam. The miniature TEPC simulated 1 μm of tissue. The lineal energy dose distributions were measured at four different points, *A* to *D*, along the depth-dose curve. The depth is marked on the upper horizontal axis and the relative dose on the right vertical axis. (From De Nardo L. et al., *Radiat. Prot. Dosim.* 110(1/4), 681–686 (2004).)

clinically interesting value is the product of RBE_{μ} and the relative absorbed dose along the depth absorbed dose curve. This product increased by about 10% along the SOBP. Although the $r(y)$ function was derived from measurements at 2 μm it was in this instance applied to results determined at 1 μm. Differences in results between dose distributions for the two volumes have been found to be small.

4.5.3 The Response Function $r(y)$ Applied to a Proton Therapy Beam of 85 MeV

The same response function $r(y)$ was also used by Wambersie et al. (2002) to judge the possible increase in RBE in a proton beam of 85 MeV in particular at the distal end of the SOBP (Figure 4.12). Besides the $r(y)$ function the figure includes single-event dose distributions observed in a ^{60}Co γ beam as well as those observed in the unmodulated and modulated proton beam. The proton single-event dose distributions as compared to the γ dose distribution have a larger dose fraction in the y-interval where

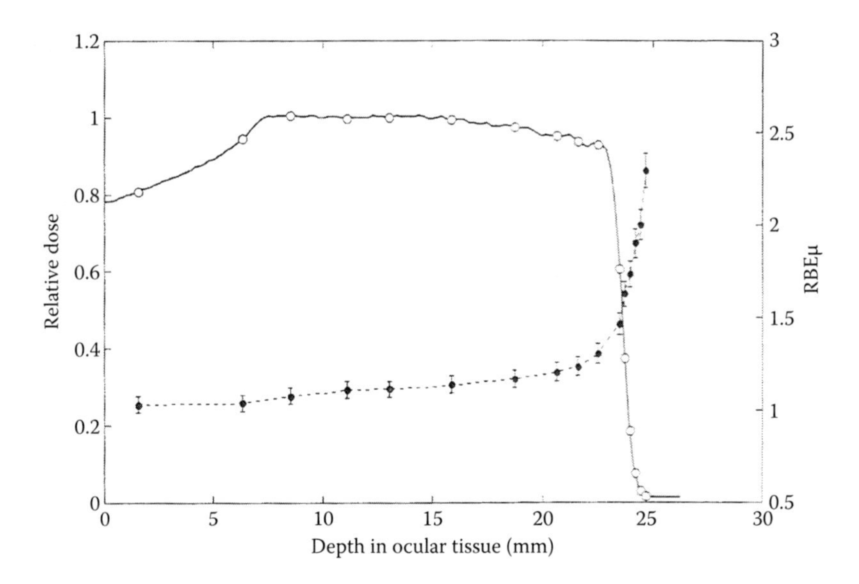

FIGURE 4.14 The dotted line and full circles are RBE values (right-hand vertical axis) calculated from Equation 4.9 using the response function for early effects in mice (Figure 4.12) and the lineal energy dose distributions shown in Figure 4.13. The full line and open circles is the depth-dose curve shown in Figure 4.13. The total uncertainty at all points corresponding to one relative standard deviation was 4.1%. (From De Nardo L. et al., *Radiat. Prot. Dosim.* 110(1/4), 681–686 (2004).)

the $r(y)$ function begins to increase. However, as only small parts of the proton dose distributions are found in the y-interval, where the $r(y)$ function is varying rapidly around its maximum, the authors concluded that the RBE of this proton beam will be close to that of ^{60}Co γ. This conclusion was substantiated by biological experiments using rat crypt cells (Gueulette et al., 1996).

4.5.4 The Response Function $r(y)$ Applied to a ^{12}C Ion Therapy Beam of 194 MeV u^{-1}

A similar investigation for carbon ion beams was reported by Gerlach et al. (2002). Single-event dose distributions were determined experimentally with an elongated cylindrical TEPC with 3 mm diameter and 30 mm of length. The detector simulated a volume with 1 μm diameter. The measurements were performed in a Lucite phantom and in carbon ion beams of several energies. Figure 4.15a and b show the dose distributions as function of the depth along the Bragg peak plateau and around the Bragg peak respectively in a ^{12}C ion beam of 194 MeV u^{-1}. When the same

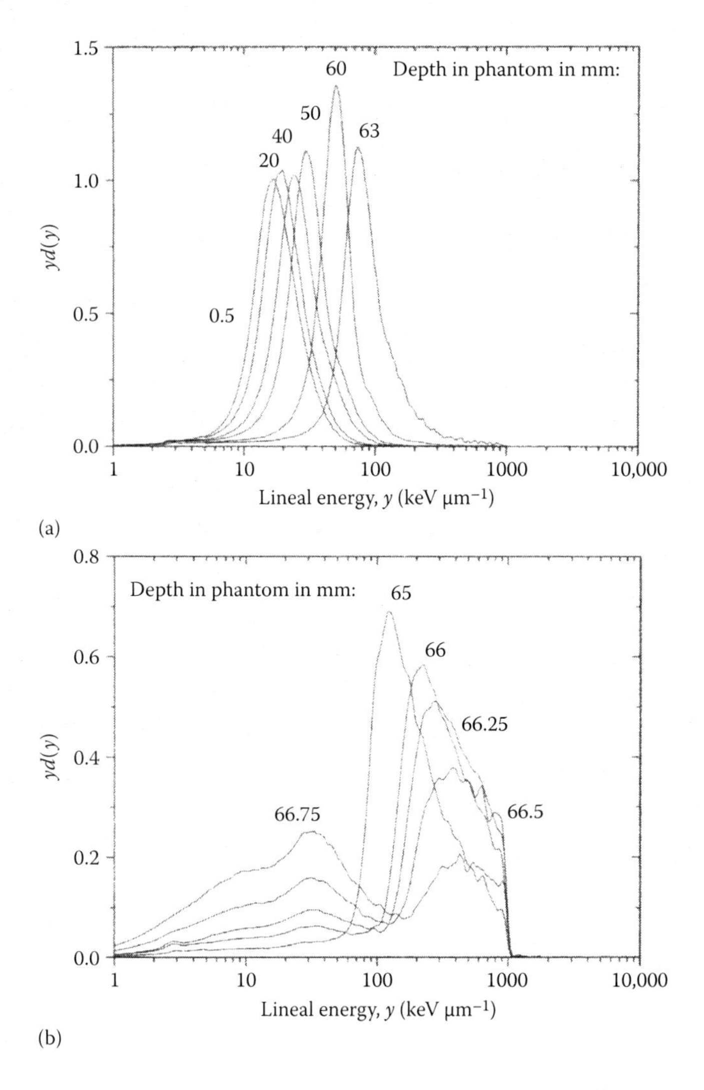

FIGURE 4.15 (a) Dose distributions of lineal energy in a ^{12}C ion beam of 194 MeV u^{-1} measured at different depths in the plateau region. The simulated volume was cylindrical and the diameter simulated 1 μm. (b) Dose distributions of lineal energy measured around the Bragg peak for the same beam as in (a). (From Gerlach R. et al., *Radiat. Prot. Dosim.* 99, 413–418 (2002).)

$r(y)$ function as used in the preceding text was applied to the distributions, the derived R-value changed from 1.7 at depth zero to 3.6 at 62 mm depth close to the end of the Bragg peak. In Figure 4.16 the estimated RBE for several different energies of the carbon ion beam and at 5 mm depth in Perspex are shown. The R calculated for 290 MeV u^{-1} is approximately 1.3.

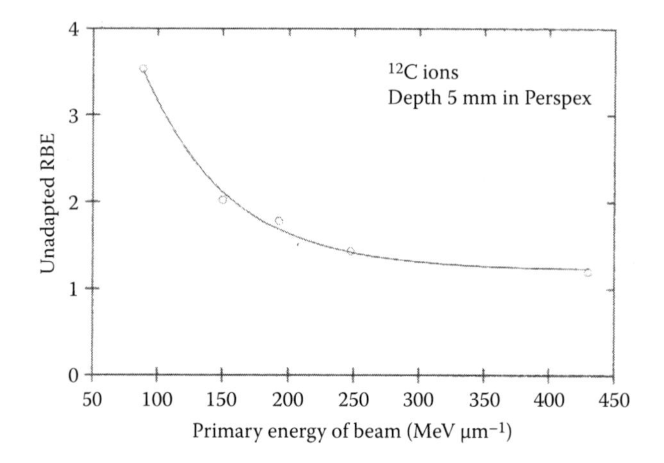

FIGURE 4.16 Estimated RBE values at different primary ^{12}C ion beam energies. The values have been calculated with the neutron therapy algorithm and the weighting function for neutrons. (From Gerlach R. et al., *Radiat. Prot. Dosim.* 99, 413–418 (2002).)

This value may be compared to the value of 1.1 reported by Hultquist (Section 4.6.3 of this chapter) from ratios of dose mean lineal energy values in a 10-nm site.

4.5.5 The Response Function *r*(*y*) Applied to a Neutron Beam Designed for BNCT

The purpose of this study was to determine the quality at different depths of a neutron beam designed specifically for BNCT. Endo et al. (2004) used two different proportional counters to measure the various dose components. In this study a commercial TEPC (LET 1/2 inch, Far West Technology) filled with methane-based TE gas simulating a 1-μm sphere was used as well as another geometrically identical detector furnished with carbon walls (LET 1/2 inch, Far West Technology). The carbon-walled detector was filled with CO_2 gas. This detector, which was designed to contain no hydrogen, allowed the measurement of recoil carbon ions from neutron elastic scattering and alpha particles from the $^{12}C(n, \alpha)$ reaction in tissue. The microdosimetric analysis commenced with the subtraction of the gamma component of the dose determined by the carbon-walled counter from both the single-event distributions measured by the tissue-equivalent counter and the carbon counter. Although the precise method of executing this gamma dose subtraction is not given by the authors, such methods are described in Chapter 5, Section 5.8.1. The resulting event-size spectrum of the carbon-walled counter was then scaled by a factor of 0.39 to match the number of carbon atoms in the A-150 plastic-walled TEPC. Following this adjustment, the carbon recoil events in the two spectra were expected to be equal; however, there was a clear difference between the two counters in this region of the event-size spectrum. This difference was attributed to alpha recoils from $^{17}O(n, \alpha)$ reaction in the A-150 wall of the TEPC. The peak observed in the carbon counter distribution in the same region of the proton peak in the TEPC at around 20 to 100 keV μm^{-1} was unexpected and was suggested to be due to (n, p) reactions in impurities in the carbon counter. From this analysis, dose fractions for the neutron beam from alpha particles, protons and carbon ions were derived, yielding values of 90%, 7% and 3% respectively. Measurements were performed at different depths in a phantom, and when the $r(y)$ response function of Tilikidis et al. (1996) was applied to the TEPC event-size spectrum a change in expected RBE from 3.6 at 5 mm to 2.9 at 41 mm depth was obtained (Figure 4.17). In keeping with the motivation of this study, Endo noted that the predicted RBEs for the neutron beam were lower than the 4.7 value expected for the alpha particles and lithium recoil nuclei produced by boron capture, and the decreasing trend of the beam RBE with depth was a potential benefit for normal cells with no boron uptake.

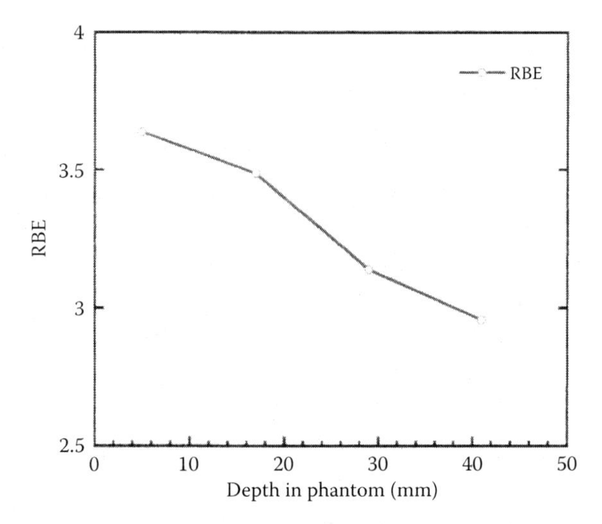

FIGURE 4.17 The RBE estimated for the clinical radiation field at KUR. (From Endo S. et al., *Radiat. Prot. Dosim.* 110(1/4), 641–664 (2004).)

4.6 MODELS FOR THE BIOLOGICAL ACTION OF IONISING RADIATION

In this section, a few models for predicting RBE that include both physical and biological terms and involve experimental microdosimetry combined with biological observations are reviewed. The models are under constant development, and for a comprehensive discussion the reader should, of course, consult the original references, especially when it comes to detailed biological mechanisms. In this section, we focus on two classes of models, summarising their basic principles and their applicability for high-LET radiation therapy:

1. The Theory of Dual Radiation Action (TDRA), a model for calculating RBE from ratios of dose mean lineal energy for nanometre volumes as well as the microdosimetric-kinetic models (Sections 4.6.2, 4.6.3 and 4.6.4)
2. The amorphous track-structure and the local effect models (Sections 4.6.5 and 4.6.6)

However, before descriptions of these specific models are given, the variance of the microdosimetric distributions will be dealt with, as this affects the precision with which the energy imparted can be known independent of the experimental uncertainties.

4.6.1 The Statistical Uncertainty due to Fluctuations in Energy Deposition

In Section 4.2 of this chapter it was mentioned that differences of 5% to 10% in absorbed dose might be clinically detected in the outcome of a treatment. Before involving biological models it is worthwhile to recall the inherent uncertainty in the deposition of the energy in small volumes. The mean specific energy, $\bar{z}$, equals the absorbed dose, D, at large dose values. Its relative variance, $V_{r,D}$, is

$$V_{r,D} = \frac{\bar{z}_{D,s}}{D} \tag{4.9}$$

where $\bar{z}_{D,s}$ is the mean specific energy in a single energy deposition event. Tumour cells are dependent on their microenvironment as well as on signalling from cells in their vicinity (Hanahan and Weinberg, 2011), and the dose to a volume larger than a cell nucleus may be relevant. Nevertheless, the goal is usually to eliminate tumour cells, and if they incur sufficient damage they will suffer proliferative death. If we assume that the absorbed dose to the cell nucleus is critical we can investigate the influence of the microdosimetric relative variance on the uncertainty of the absorbed dose.

If a total absorbed dose of D = 60 Gy of ^{60}Co γ radiation is given to a treatment volume, the microdosimetric uncertainty in the deposition of the energy in a cell nucleus with 8 μm diameter, calculated as relative standard deviation $\sigma_{r,D} = \sqrt{V_{r,D}}$ in a volume, is 0.65% since $\bar{z}_{D,s}$ = 2.5 mGy (derived from Lillhök et al., 2007b). Consider the same volume but now exposed to a neutron therapy beam. If the total absorbed dose is D = 22 Gy, $\sigma_{r,D}$ = 9.2% if $\bar{z}_{D,s}$ = 188 mGy (measured in the former neutron beam at Louvain la Neuve; Lillhök et al., 2007a). This uncertainty has the meaning that 67% of all cell nuclei are given a dose in the interval $D - \sigma$ to $D + \sigma$. In the low-LET beam $\sigma_{r,D}$ is within the accepted uncertainty of ±3%, but ±9% as found in the neutron beam may be limiting a successful treatment. From a microdosimetry point of view, a way to reduce this uncertainty is to add a dose fraction of low-LET radiation. If the relative fraction of the dose due to γ-rays is w_γ, the combined dose mean specific energy is

$$\bar{z}_{D,s} = \bar{z}_{D,s,\gamma} w_\gamma + \bar{z}_{D,s,n}\left(1 - w_\gamma\right) \tag{4.10}$$

and the uncertainty is reduced as

$$\sigma_{r,D} = \sqrt{\frac{\bar{z}_{D,s}}{D}} = \sqrt{\frac{\bar{z}_{D,s,\gamma} w_\gamma + \bar{z}_{D,s,n}\left(1 - w_\gamma\right)}{D}} \tag{4.11}$$

If the total absorbed dose is kept 22 Gy, but the gamma dose fraction is made w_γ = 70% (15 Gy), the neutron dose then becomes 7 Gy and $\sigma_{r,D}$ = 5%. If instead the gamma dose is multiplied by 3.2 to compensate for a difference in radiation equality between neutrons and γ-rays, the gamma dose is now 48 Gy and the total dose is 55 Gy. The relative standard deviation for the beam now becomes $\sigma_{r,D}$ = 2.1%. This calculation is made only as an illustration. In practice, several other uncertainties would have to be taken into account. The uncertainty in absorbed dose to the cell nuclei is also dependent on their volume.

The influence of this uncertainty on the tumour control probability was discussed by Lindborg and Brahme (1990). They suggested that a total dosimetric uncertainty should include not only the traditional dosimetric uncertainty but also the microdosimetric uncertainty, which then would affect the tumour control probability.

4.6.2 Theory of Dual Radiation Action

An early microdosimetric model for describing the relation between early effects and microdosimetry was the Theory of Dual Radiation Action (TDRA) by Kellerer and Rossi (1972). The TDRA was not meant to be directly applicable to radiation therapy, but it has certainly had an impact on these discussions. The TDRA has gone through a number of changes, and an excellent review is found in the textbook by Rossi and Zaider (1996). In the original theory, the so-called *site model*, it was assumed that an initial lesion responsible for radiation effects is produced by combinations of pairs of sub-lesions produced at a rate proportional to the absorbed dose. At high LET, pairs of combining sub-lesions are produced most probably by single events, whereas for low LET, they are produced most probably by two separate and independent events. The sub-lesions were thought to be created in a volume of a certain size and the geometric distribution of the sub-lesions within this volume was not considered important. The biological effect (*BE*) in the low- and high-LET beams was related to the absorbed dose in the two beams according to

$$BE(D_{\mathrm{L}}) = c_{\mathrm{L}}\left(\bar{z}_{D,s,\mathrm{L}} D_{\mathrm{L}} + D_{\mathrm{L}}^{2}\right) \tag{4.12}$$

$$BE(D_{\mathrm{H}}) = c_{\mathrm{H}}\left(\bar{z}_{D,s,\mathrm{H}} D_{\mathrm{H}} + D_{\mathrm{H}}^{2}\right) \tag{4.13}$$

When the dose of low-LET radiation (D_{L}) and the dose of high-LET radiation (D_{H}) produce the same level of *BE*, an expression for the RBE may be derived from

$$c_{\mathrm{L}}\left(\bar{z}_{D,s,\mathrm{L}} D_{\mathrm{L}} + D_{\mathrm{L}}^{2}\right) = c_{\mathrm{H}}\left(\bar{z}_{D,s,\mathrm{H}} D_{\mathrm{H}} + D_{\mathrm{H}}^{2}\right) \tag{4.14}$$

At very small absorbed dose values where the D^2 term can be ignored and on condition that $c_{\mathrm{L}} = c_{\mathrm{H}}$, the RBE becomes equal to the ratio of the dose mean specific energies (or dose mean lineal energies) for the two radiation qualities.

$$\mathrm{RBE} = \frac{D_{\mathrm{L}}}{D_{\mathrm{H}}} \approx \frac{\bar{z}_{D,s,\mathrm{H}}}{\bar{z}_{D,s,\mathrm{L}}} = \frac{\bar{y}_{D,s,\mathrm{H}}}{\bar{y}_{D,s,\mathrm{L}}} \tag{4.15}$$

Ratios of the dose mean lineal energy determined in a volume with a diameter in the range 1 to 3 μm were found to be close to observed RBE values for several biological endpoints.

The saturation effect mentioned in Section 4.3 (Figure 4.2) was taken into account by replacing $\bar{z}_{D,1H}$ in Equation 4.15 with

$$z_D^* = z_0^2 \int_0^\infty \left(1 - e^{-\left(\frac{z}{z_0}\right)^2}\right) f_1(z)dz/\bar{z}_s \tag{4.16}$$

As z changes drastically with volume a lineal energy value y^* corresponding to z_D^* was introduced and

$$y^* = \frac{y_0^2 \int_0^\infty \left[1 - e^{-\left(\frac{y}{y_0}\right)^2}\right] f(y)dy}{\int_0^\infty yf(y)dy} \tag{4.17}$$

The value of y_0 is typically in the range 125 keV μm^{-1} to 150 keV μm^{-1}.

Following the publication of the TDRA, a series of radiobiological experiments were made with ultra-soft x-rays with energies below 0.5 keV (for instance, Goodhead et al., 1979). With the range of the photoelectrons produced by these ultra-soft x-rays being 17 nm or less, this indicated a critical volume much smaller than the volume size suggested by the TDRA.

In a revised version of the TDRA called the *distance model*, the distances between the sub-lesions were taken into account. The theory includes a geometrical distribution of the sub-lesions, which requires that the points where energy is transferred is defined, as well as information on the structure in which sub-lesions are formed. The revised TDRA still requires that two sub-lesions are needed to generate a lethal lesion. The probability that two energy transfer events a distance x apart will combine to form a lesion is called $\gamma(x)$ and was defined as the product of a structure function $s(x)$ and a sub-lesion combination probability function $g(x)$ with

$$\gamma(x) = \frac{g(x)s(x)}{4\pi \varrho x^2 \int_0^\infty g(x)s(x)dx} \tag{4.18}$$

where ϱ is the density of the matter. The yield of lesions $\varepsilon(D)$ is given by

$$\varepsilon(D) = K\left[D\int_0^\infty t(x)\gamma(x)dx + D\right] \tag{4.19}$$

Here $t(x)$ is the proximity function for transfer points belonging to the same track and is dependent on radiation quality. The $\gamma(x)$ function is not well known but an attempt to estimate it was made in the molecular ion beam experiment mentioned in Section 4.3.2. The result showed that sub-lesions interact mostly over short distances of less than about 20 nm, but interactions may occur up to a few micrometres.

Following this work, the TDRA was developed further into what is called the Compound Dual Radiation Action in which an attempt was made to identify the molecular nature of cytological damage previously termed sub-lesions and lesions. Examples of such elements are DSBs and single-strand breaks (SSBs). SSBs are generated at a rate proportional to the specific energy, z. They combine pairwise into DSBs at a rate proportional to the square of the specific energy, z^2, in a site that is so small that the geometric distribution of relevant transfer points within the site may be regarded as unimportant.

4.6.3 A Model for Estimating Weighting Factors from $\bar{y}_D$ Values

A model for deriving radiation weighting factors from ratios of $\bar{y}_D$ values has been suggested by Lindborg and Grindborg (1997). This model also assumes that there is a site within which the geometrical details of the energy transfer point can be disregarded. It is further assumed that the time between dose fractions is sufficient for repair mechanisms to eliminate the repairable lesions and that the remaining lesions are complex and difficult to repair or are irreparable and that the α coefficient in the linear quadratic expression for dose effects reflects those severe lesions. As a first approximation α is considered proportional to $\bar{y}_D$ for a particular volume size, which is the same for all radiation qualities. The hypothesis was formulated as

$$\frac{\bar{y}_{D,b}}{\bar{y}_{D,\gamma}} = \frac{\alpha_b}{\alpha_\gamma} \tag{4.20}$$

In Equation 4.5 (Section 4.1), if $d_\gamma = 2$ Gy and if for early reacting tissues and many tumours α/β is set equal to 10 for low-LET beams including protons and if this ratio is assumed to be so much larger in ion beams, that the ratio $d_b/(\alpha/\beta)_b$ can be ignored, then a relation between clinical RBE values (W_b values) and $\bar{y}_D$ can be derived. For an ion-beam relative to a gamma beam Equation 4.5 becomes

$$W_{\text{ion}} = \frac{\alpha_{\text{ion}}}{\alpha_\gamma(1 + 0.2)} \tag{4.21}$$

Thus ratios of α values can be derived from Equation 4.5. For ions, neutrons, protons and x-rays the ratios become

$$\frac{\alpha_{\text{ion}}}{\alpha_\gamma} = 1.2W_{\text{ion}} \tag{4.22}$$

$$\frac{\alpha_n}{\alpha_\gamma} = 1.2W_n \tag{4.23}$$

$$\frac{\alpha_p}{\alpha_\gamma} = W_p \tag{4.24}$$

$$\frac{\alpha_x}{\alpha_\gamma} = W_x \tag{4.25}$$

The numerical values for the α ratios are obtained if the W_b values from Table 4.1 are inserted into Equations 4.22 to 4.25. These α ratios, which are independent of volume size, were then compared to $\bar{y}_D$ ratios, which vary with the volume size. $\bar{y}_D$ ratios for volumes with diameters from a few nanometres up to several micrometres were determined for the therapy beams listed in Table 4.1. The condition expressed by Equation 4.20 was found to be fulfilled for all the investigated radiation qualities for volumes with diameters in the interval 10 nm to 15 nm. Figure 4.18 shows the resulting approximately linear relation found between the α ratios and $\bar{y}_D$ ratios. The $\bar{y}_D$ values were partly determined from variance–covariance measurements and partly from Monte Carlo track-structure calculations in combination with condensed Monte Carlo calculations (Hultqvist et al., 2010; Lillhök et al., 2007a; Lindborg et al., 2013).

The capacity of $\bar{y}_D$ ratios to predict W_b values was tested using Equations 4.22 to 4.25 with $\bar{y}_D$ ratios for a cylinder with 10 nm diameter and 10 nm height replacing the α ratios. The values calculated in this way, called W_{calc}, are shown in Table 4.4 along with clinically reported W_b values.

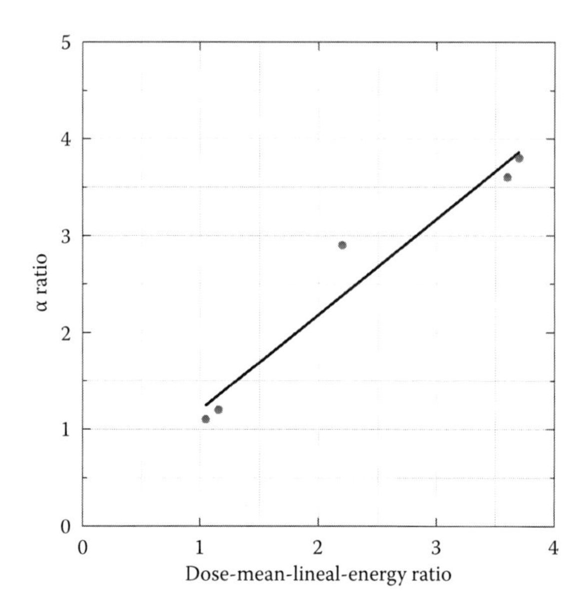

FIGURE 4.18 The α ratio calculated from clinically observed RBE values in different radiation therapy beams (Equation 4.20), with ^{60}Co γ as reference radiation. The α ratio is presented as function of the $\bar{y}_D$ ratio for the radiation beams for which the RBE values were determined and for cylindrical volumes with 10 nm diameter and height. The points represent from bottom left: 100 kV x-rays, 175-MeV protons in the SOBP plateau, ^{12}C ions 290 MeV u^{-1} in the SOBP plateau, the distal end of the SOBP and a neutron therapy beam. (From Lindborg L. et al., *Phys. Med. Biol.* 58, 3089–3105 (2013).)

TABLE 4.4 Radiation Weighting Factors W_{calc} Calculated According to Equations 4.22 to 4.25 with $\bar{y}_D$ Ratios Replacing the α Ratios

Radiation Quality	W_b	W_{calc}
X-ray/^{60}Co γ	1.25	1.2
p (175 MeV)/^{60}Co γ	1.1	1.1
in SOBP at 5 cm	IAEA (2008)	
^{12}C (300 MeV u^{-1})/^{60}Co γ	2.4	1.8
Centre of SOBP	Matsufuji et al. (2007)	
^{12}C (300 MeV u^{-1})/^{60}Co γ	3.0	2.8
Distal end of SOBP	Matsufuji et al. (2007)	
Neutrons/^{60}Co γ	3.2	3.1
	Batterman et al. (1981)	

Note: $\bar{y}_D$ determined for a cylindrical volume with 10 nm diameter and height except for the reference ^{60}Co γ beam, where the volume was a sphere with the simulated diameter of 10 nm. The reference $\bar{y}_D$ value was 16.6 keV μm^{-1}. Clinically observed radiation weighting factors W_b from Table 4.1 are included for comparison. The therapy beams are the same as above.

The agreement between W_b and W_{calc} is within 10% except for the centre of the plateau, which may be related to a lack of sufficient information about the filter in the clinical beam.

A further test of the model was reported by Hultqvist (2011). The change of W_{calc} with depth was calculated for a depth absorbed dose profile in a ^{12}C ion beam (290 MeV u^{-1}). $W_{calc,ion}$ changed from approximately 1.1 at the entrance to approximately 3.5 at the distal end of the Bragg peak. Within the SOBP itself the change was from 1.5 to 3.5 (Figure 4.19). These results are in good agreement with what has been reported for the carbon ion beam at HIMAC, Chiba (Endo et al., 2010; Matsufuji et al., 2007) and the results reported in Section 4.5.4 The observation supports the idea that characterising therapy beams at a simulated volume of about 10 nm diameter is meaningful at least for early reacting tissues.

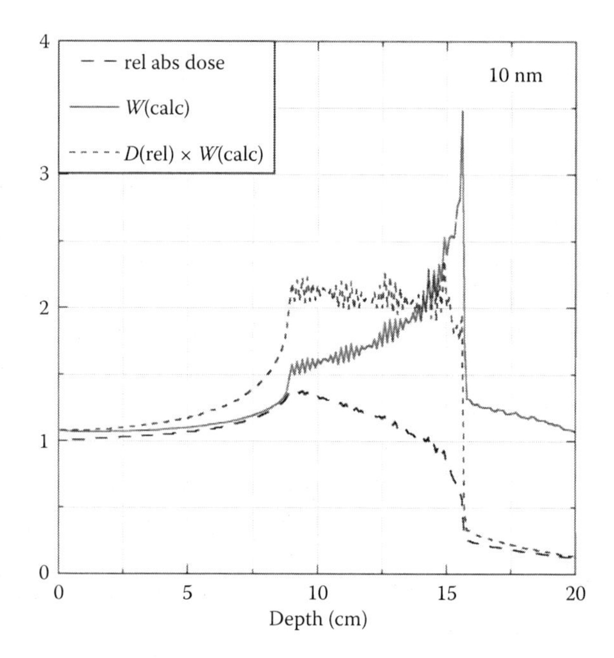

FIGURE 4.19 **(See colour insert.)** Depth profiles in water of a 6 cm wide ^{12}C ion beam, 290 MeV u^{-1}. The long dashed line is the absorbed dose profile; the solid line is the calculated weighting factor, W_{calc} (Equation 4.8) with $\bar{y}_D$ values determined for a cylinder of 10 nm height and diameter. The short dashed line is the product of the two other curves and corresponds to the expected biological efficiency in this model. This curve is fairly constant over the SOBP region. (From Hultqvist M. *Secondary absorbed close distribution and radiation quality in light ion therapy* PhD thesis, Department of Physics, Stockholm University (2011).)

Although some of the lineal energy measurements have been made in volumes where the uncertainty is large (as discussed in Chapter 3), it is the ratio that enters in this derivation of the relation to clinically observed effects. Some of the systematic uncertainties in absolute values obtained through measurements in the nanometre range are expected to be cancelled or at least become smaller when relative values, such as $\bar{y}_D$ ratios, are used. It may also be said that when measured ratios of $\bar{y}_D$ are used in practice the ratio is close to the ratio of the corresponding mean value of the single-event ionisation distributions for the two radiation qualities.

4.6.4 The Microdosimetric-Kinetic Model

In the 1990s a model was developed and published called the microdosimetric-kinetic (MK) model (Hawkins, 1994, 1996, 1998, 2003, 2006; Hawkins and Inaniwa, 2014). The model provides a quantitative explanation for the relation between RBE and LET for reproductive cell death in mammalian cells. It has many similarities with the earlier model of the TDRA. The short description given here is based mostly on the account given by Hawkins in 2003.

In the MK model sub-volumes of the cell nucleus called domains are defined and the distribution of specific energy in the different domains is directly related to the creation of lesions. In particular, the dose mean specific energy in a domain is the important microdosimetric parameter. A variation of the DNA content among the domains is another model parameter. The MK model postulates the formation of two types of lesion, type I and type II. Both are proportional to the specific energy in a domain and to the mass of the DNA content in that domain. Whereas type I lesions are immediately lethal, type II lesions may undergo one of four transformations and may or may not become lethal. A lesion of type I is said to be either a special unrepairable form of a DSB or some other unrepairable lesion (Hawkins, 1996). A type II lesion is likely a DSB. Lethal lesions are assumed to be distributed randomly and individually with equal probability among the subsets of domains that absorb exactly the same dose. The size of the domain can be experimentally determined from results of RBE experiments at a very low absorbed dose (called RBE_1) in track segment experiments using RBE_1 plotted as a function of LET (Hawkins, 1998). Using this approach, the range of domain diameters was found to vary from 0.5 μm to 1 μm. The MK model includes two differential equations (differential with respect to time) that describe the kinetics of the formation of lesions.

In the MK model and for LET radiations low enough that the number of lethal lesions among the cell population is Poisson distributed, the average number of lethal lesions per cell ($\bar{w}_n$), n standing for the sensitive nuclear volume, is

$$\bar{w}_n = [\alpha_0 + \beta(\bar{z}_{D,s})_d]D + \beta D^2 = \alpha_P D + \beta D^2 \tag{4.26}$$

α_0 and β are related to the parameters of the kinematics of the formation of lesions as well as to the content of DNA in the domain of diameter d and $(\bar{z}_{D,s})_d$ is the dose mean specific energy of the single-event distribution in a domain. When D is small and if lethal lesions are Poisson distributed independently of radiation quality, RBE_1, now called $\mathrm{RBE}_{1\mathrm{P}}$ (P stands for Poisson distribution), is

$$\mathrm{RBE}_{1\mathrm{P}} = \frac{\alpha_0 + \beta(\bar{z}_{D,s})_d}{\alpha_\mathrm{R}} = \frac{\alpha_\mathrm{P}}{\alpha_\mathrm{R}} \tag{4.27}$$

α_R is the initial slope of the survival curve for the reference low-LET radiation.

A consequence of Equation 4.27 is that $\mathrm{RBE}_{1\mathrm{P}}$ will increase linearly with $(\bar{z}_{D,s})_d$, if α_0 and β are independent of radiation quality. Such linearity is not in agreement with experiments. This is interpreted as if lethal lesions are no longer Poisson distributed among the cell nuclei for radiations of high LET and a saturation correction factor is introduced. If D is low enough that the probability a cell is hit by more than one event can be ignored, α_P in Equation 4.27 is replaced by

$$\alpha^* \approx \left(\frac{1 - e^{-\alpha_\mathrm{P}(\bar{z}_{D,s})_n}}{\alpha_\mathrm{P}(\bar{z}_{D,s})_n}\right) a_\mathrm{P} = \left(\frac{1 - e^{-\alpha_\mathrm{P}(\bar{z}_{D,s})_n}}{\alpha_\mathrm{P}(\bar{z}_{D,s})_n}\right) [\alpha_0 + (\bar{z}_{D,s})_d\beta] \tag{4.28}$$

Here $(\bar{z}_{D,s})_n$ is the dose mean specific energy to the sensitive nuclear volume, which is defined as the combined volume of domains that contains the whole distribution of the DNA. The sensitive nuclear volume is less well defined but is expected to be about 100 times the volume of a domain. If α^* is inserted into Equation 4.26 instead of α_P, the equation applies to radiation of any LET. When the absorbed dose is small, the average number of lethal events is

$$\bar{w}_n = \alpha^* D \approx \left(\frac{1 - e^{-\alpha_P (\bar{z}_{D,s})_n}}{\alpha_P (\bar{z}_{D,s})_n} \right) [\alpha_0 + \beta(\bar{z}_{D,s})_d] D \tag{4.29}$$

Thus the number of lethal lesions is dependent on the $\bar{z}_{D,s}$ in both the domain and the nucleus.

An example of how to apply the 1996 version of the MK model to cell survival experiments has been reported by Kase et al. (2006). They applied the MK model to irradiations of both human salivary gland tumour cells as well as a normal fibroblast cell line, GM0538 cells. Irradiations were carried out with ^{60}Co γ rays and various ion beams, both in the Bragg peaks of mono-energetic beams and in the SOBP of modulated beams, covering a wide range of LET from 0.5 to 880 keV μm^{-1}. Single-event dose distributions were determined with a commercial TEPC simulating a microscopic domain of 1 μm and the mean values $\bar{y}$ and $\bar{y}_D$ evaluated for the same locations as the cells were irradiated. The authors argue that a saturation correction should be applied not only to the dose mean value $(\bar{z}_{D,s})_n$, as suggested by Hawkins (Equation 4.29), but also to the full measured lineal energy distribution; consequently they used a modified correction, α^*, such that

$$\bar{w}_n = \left[\alpha_0 + \beta\left(z^*_{D,S}\right)_d\right] D + \beta D^2 \tag{4.30}$$

with

$$\left(z^*_{D,S}\right)_d = \frac{\ell}{m} y^* = \frac{\bar{\ell} y_0^2 \int \left[1 - e^{-\frac{y^2}{y_0^2}}\right] y f(y) dy}{m \int y f(y) dy} \tag{4.31}$$

where y_0 = 150 keV μm^{-1}.

Kase et al. concluded that a better correlation was now observed between experimentally determined α values and y^* than in the original publication by Hawkins. Exceptions being the results observed in the ^{60}Co γ as well as ion beams with L > 450 keV μm^{-1}. The MK model has been included in the PHITS code by the code developers (Sato et al., 2011), who also show with illustrative calculations that the experimentally determinable microdosimetric parameter y^* removes the ion-species dependency observed in the relationship between α and LET.

4.6.5 Amorphous Track Structure Models

About the same time the TDRA was being advanced, an alternative model was being developed to predict the biological effects of charged particle irradiation. An essential feature of this alternative model was that there was no principal difference between the biological action of low- and high-LET radiation because for both radiation qualities the effect is brought about through the interaction of electrons and it matters not whether the electron is generated by a photon or by a charged particle; only the mean concentration of these electrons determines the effect. A second feature of the model was therefore that the track structure of a charged particle could be adequately represented by the radial-dose profile around the track and that the stochastic nature of secondary electron emission could be ignored. For this reason, the model became known as the 'amorphous track-structure model'. Bringing these two aspects of radiation interaction together, it was proposed that the response of a biological system such as a cell to charged particles could be predicted by combining the knowledge of the microscopic radial dose distribution around a charged particle track and the γ-ray dose–response curve for the cell or biological system under test. The amorphous track model was developed mainly by Robert Katz and coworkers in the late 1960s and early 1970s (Butts and Katz, 1967; Katz et al., 1972) and proved successful for the prediction of charged particle RBE for viruses, bacteria and mammalian cells. A recent review on the principles of Katz's cellular track-structure model was presented at the 16th International Symposium on Microdosimetry (Waligórski et al., 2015).

Since the radial dose distribution around a charged particle track is determined by the range of the secondary electrons emitted from the track, and these will be on the scale of nanometres to micrometres, amorphous track models played an important role in the evolution of microdosimetry and are still an integral part of the discipline. However, there are no parameters in the amorphous track-structure model that deal directly with the stochastic nature of energy deposition. Although there is no direct link between amorphous track models of biological action and experimental microdosimetry, amorphous track-structure models have developed significantly over the years and are playing an important role in treatment planning for charged particle therapy using different ion beams compared or combined with more conventional x- and γ-ray modalities. It is therefore valuable to review briefly the latest development of this class of model known as the local effect model (LEM).

4.6.6 The Local Effect Model

In 1994 a model for calculation of cell inactivation by heavy ions was published (Scholz and Kraft, 1994). The model assumes that cell inactivation is the result of lethal events caused by a local modification of a sensitive target (DNA) in a small sub-volume, V, of the cell nucleus; the lethal lesions are assumed to be all of the same type independent of radiation quality. No interactions of sub-lethal damages are supposed to occur over micrometre distances.

The lethal event density, ν, in a sub-volume, V, of a cell nucleus is supposed to be proportional to the local dose, d, in this volume. An essential assumption in this theory is that ν is independent of the radiation quality. Thus

$$\nu_{\text{ion}}(d) = \nu_x(d) \tag{4.32}$$

ν_{ion} and ν_x are the lethal event densities observed in an ion beam and x-ray beam respectively. The local dose, d, is the expectation value of the specific energy in V. Although a sub-volume is critical for the calculation of the local dose, the whole cell nucleus is regarded as radiosensitive. The relationship between the surviving fraction of cells after irradiations with x-rays and ions is as follows: with x-rays a lethal event density, $\nu_x(d)$, is created that is assumed to be the same over the whole cell nucleus. If $\bar{N}_x(D)$ is the expectation number of lethal events in the cell with the cell nucleus volume, V_{nucl}, at the absorbed dose, D, then

$$\nu_x(d) = \nu_x(D) = \frac{\bar{N}_x(D)}{V_{\text{nucl}}} \tag{4.33}$$

The probability a cell will survive at the absorbed dose, D, after irradiation with x-rays is assumed to be predicted by Poisson statistics and

$$S_x = e^{-\bar{N}_x(D)} \tag{4.34}$$

and

$$\nu_x(D) = \frac{-\ln S_x}{V_{\text{nucl}}} \tag{4.35}$$

The cell survival curve after x-ray irradiation is given by

$$-\ln S_x = \alpha_x D + \beta_x D^2 \quad \text{for } D < D_t \tag{4.36}$$

$$-\ln S_x = \alpha_x D_t + \beta_x D_t^2 + s_{max}(D - D_t) \quad \text{for } D \geq D_t \tag{4.37}$$

where S_x is the survival fraction observed after x-ray irradiation and α_x and β_x have their usual meaning. The survival curve is thus characterised by a shoulder and a final exponential slope. D_t is a threshold dose above which the survival curve is purely exponential. Finally s_{max} is the maximum slope obtained by differentiation of the linear-quadratic survival curve at the dose where the survival becomes exponential and

$$s_{max} = \alpha_x + 2\beta_x D_t \tag{4.38}$$

After irradiation with ions the absorbed dose in each volume element of the cell is calculated according to a simplified model: the radial dose of the track core is considered constant out to a distance of 10 nm, after which it decreases as $1/r^2$, where r is the distance from the ion track core until a distance corresponding to the maximum range of the delta electrons is reached. If several ions contribute to the dose in a specific volume element, the dose is the sum of all contributions. After integration of the absorbed dose d in the volume elements over the entire nuclear volume, the expected number of lethal events in the cell is

$$\bar{N}_{ion}(D) = \int_{V_{nucl}} v_{ion}[d(x,y,z)]dV \tag{4.39}$$

As $v_{ion}(d) = v_x(d)$ it follows that

$$\bar{N}_{ion}(D) = \int_{nucl} V_x[d(x,y,z)]dV = \int_{nucl} \frac{-\ln S_x[d(x,y,z)]}{V_{nucl}} dV \tag{4.40}$$

Thus in the LEM, the relation between absorbed dose and number of lethal lesions in a cell after irradiation with ions is defined entirely by the cell survival curve as observed after x-ray irradiation and the radial dose profile along ion tracks.

The model was further explored by Scholz et al. (1997). In more recent developments of the LEM (Elsässer et al., 2010), the concept of a lethal lesion is made more sophisticated by considering biological response to be related to the initial spatial distribution of damaged DNA rather than the local dose distribution. However, in keeping with the basic concept of the

LEM, similar patterns of DNA damage are expected to lead to equal effects regardless of the radiation quality leading to the damage pattern. Reviews of the model can be found for instance in the IAEA (2008) as well as in the thesis by Laczkó (2006). LEM is used for carbon ion therapy treatment planning at GSI in Darmstadt, Germany and in other European ion-therapy centres. An attempt to involve microdosimetry in the model has been reported (Scholz et al., 1997).

In treatments of patients, efficient computerised dose planning is needed. The use of the classical quantities has facilitated this. If instead the objective is the understanding of a model, then the physics needs to be described and employing stochastic quantities can be expected to be essential.

As pointed out in Chapter 2, in nanometre volumes and in beams of high-LET radiations the main contributor to the specific energy is delta electrons. An estimate of their energy and scattering angle may be found from Equation 2.44. As ions are much heavier than electrons, the ions are not much affected by the collisions, while the electrons become scattered at large angles. The energy of electrons emitted at an angle of 75° after collisions with ^{12}C ions of 40 MeV u^{-1} and 300 MeV u^{-1} are estimated with this formula and the corresponding $\bar{y}$ and $\bar{y}_D$ values for the electrons are taken from Figure 2.11 and shown in Table 4.5. The values are here compared to values calculated for the two carbon ion energies, Figure 2.9 and from Liamsuwan et al. (2014). Both the $\bar{y}$ and $\bar{y}_D$ values for the carbon ions are close to those of electrons scattered at 75°, meaning that their dose distributions are close as well. Thus for sites of 10 nm it is reasonable to expect that delta electrons will deposit most of the energy in the ion beam, which is one of the basic assumptions in LEM. The exact $\bar{y}_D$ values for the delta electrons need a much more detailed evaluation than made

TABLE 4.5 Comparison of L, $\bar{y}$ and $\bar{y}_D$ for Carbon Ions and Delta Electrons Scattered by the Ions at 75°

	^{12}C			Delta Electron (75°)		
^{12}C ion (MeV u^{-1})	**L (keV μm^{-1})**	**$\bar{y}$ (keV μm^{-1})**	**$\bar{y}_D$ (keV μm^{-1})**	**E (keV)**	**$\bar{y}$ (keV μm^{-1})**	**$\bar{y}_D$ (keV μm^{-1})**
40	54	15	37	0.5	23	39
300	12	8.8	21	7	10	23

Note: The values of lineal energy are calculated for cylinders with 10 nm diameter and length. Values of $\bar{y}$ and $\bar{y}_D$ for carbon ions are from Liamsuwan et al. (2014) and values of $\bar{y}$ and $\bar{y}_D$ for electrons from Nikjoo and Lindborg (2010).

here. The example may illustrate that microdosimetry can contribute to the understanding of the model.

Clearly, there will be continued development of LEMs in the future, and this development will be driven by more detailed descriptions of particle track structure probably obtained through microdosimetric Monte Carlo simulations. Although it is unlikely that the classical methods of experimental microdosimetry described in this work will have a direct role to play in the further development of local effect models, unlike the case with the MK model described in Section 4.6.4, it should not be forgotten that Monte Carlo codes themselves need benchmarking and testing against experimental work designed to test their capability of calculating energy deposition on the nanometre–micrometre scale.

4.7 SUMMARY

In this chapter, we have reported on single-event dose distributions measured in several different radiation beams used in radiation therapy. Of direct interest is finding methods to determine the weighting factor needed in both radiation therapy and radiation biology that will predict the RBE. Attempts based on dose mean lineal energy in a nanometre volume as well as dose lineal energy distributions weighted with a response function for micrometre volumes give results close to the expected RBE values. In both approaches, a presumed relationship between microdosimetric quantities and RBE is needed and this relationship cannot be assumed to predict results to other cell lines or tissues than those for which they have been derived. The somewhat surprising finding, that similar results can be found from data in the nanometre and response functions determined for micrometre ranges, may arise as a consequence of single events being dominant in the causation of lethal lesions. This observation deserves further analyses and clarification.

A direct consequence of the stochastic nature of the interaction of radiation is the level of uncertainty in the energy imparted to individual cell nuclei or other targets. This uncertainty for high-LET beams may be comparable to the uncertainty that is required for curative treatments if the energy imparted to a cell nucleus of 8 μm diameter is considered relevant. The relevance of the microdosimetric uncertainty for an understanding of biological response has been reported by Navarro (2016).

At present no proven mechanistic model exists that describes all steps from energy deposition to the killing of cancer cells. With the lack of an accepted general model for going from energy imparted to biological efficiency, microdosimetric dose distributions themselves are of considerable practical value. By comparing such distributions between therapy centres, it is possible to judge the physical conditions for expecting similar treatment results.

A few examples have been described in this chapter in which microdosimetric single-event dose distributions or their mean values have been used as a descriptor of radiation quality.

A further class of biophysical models known as amorphous track-structure models were briefly introduced and discussed because of their importance in ion-beam therapy treatment planning although these models themselves do not utilise microdosimetric information in the sense of the stochastic variation of energy deposition at the micrometre or submicrometre level.

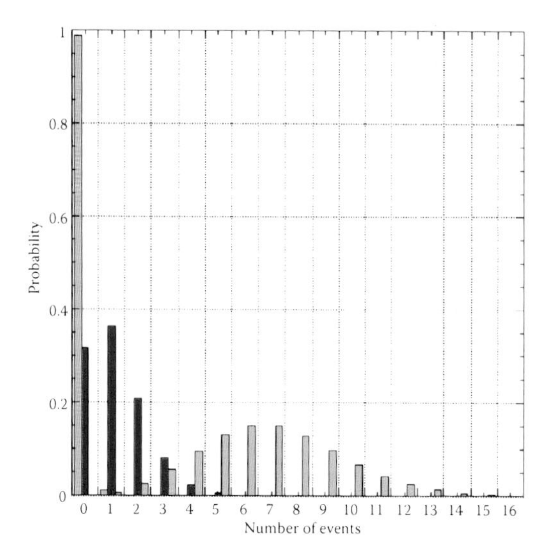

FIGURE 2.5 The probability that D is caused by ν events when D = 10 μGy (blue columns), D = 1 mGy (red columns) and D = 6 mGy (green columns) in a volume of 8 μm diameter (approximately the diameter of mammalian cell nucleus). The radiation beam is ^{60}Co γ.

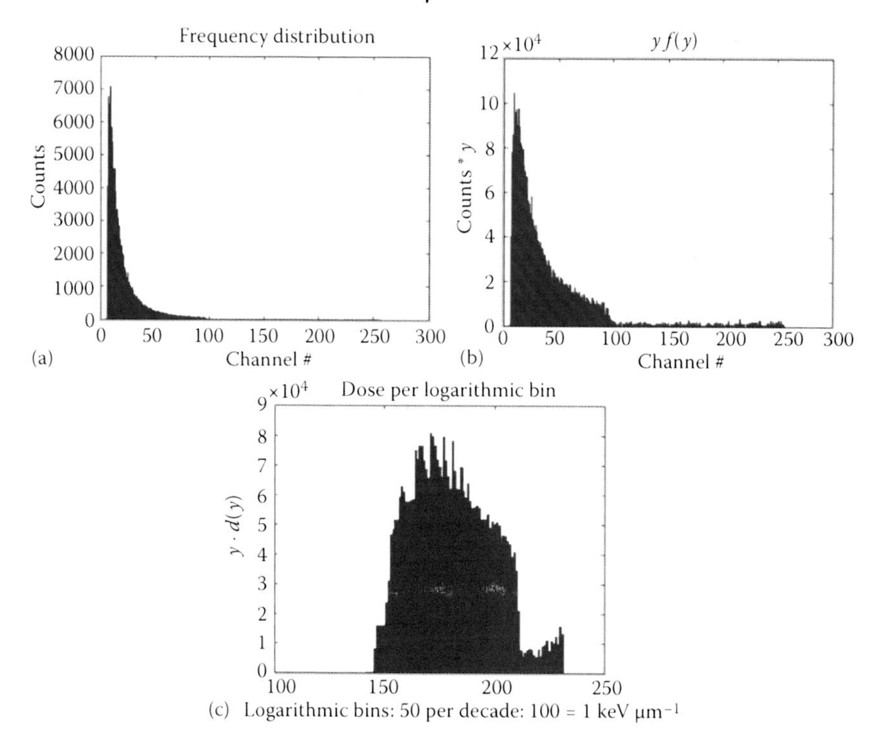

FIGURE 3.9 Data from a TEPC (REM500) instrument made in an Am–Be neutron field. The pulse height spectrum representing the frequency distribution (a) is converted into a dose distribution (b) by multiplying the frequency of an event by the size of the event. Finally, the dose distribution is redistributed on a logarithmic scale consisting of equal logarithmic intervals (c), in this case 50 logarithmic intervals per decade of lineal energy.

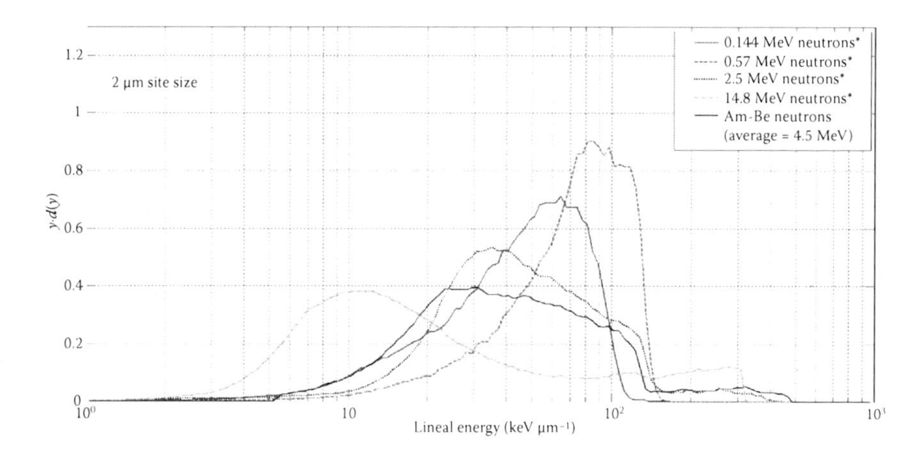

FIGURE 3.10 Single-event lineal energy dose distributions for several different neutron fields measured with a 5-inch spherical TEPC and displayed in the $yd(y)$ format on an equal logarithmic interval scale. The position of the recoil proton peak and the proton edge all convey information to the experimenter that can be useful in circumstances where the exact nature of the radiation is not known a priori.

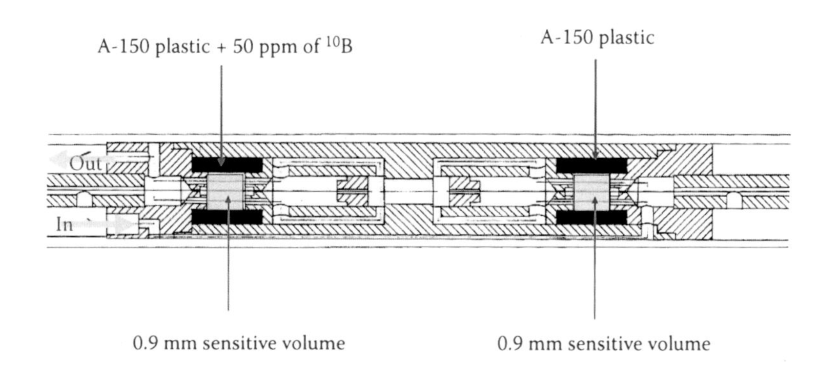

FIGURE 4.10 The twin chamber for measurements in a neutron beam for BNCT. The TEPC is a square-cylinder with a diameter and height of 0.9 mm. The A-150 wall in one of the chambers has been doped with 50 ppm of ^{10}B. The boron atoms may capture thermal neutrons and emit alpha particles along with a Li ion recoil. In BNCT the tumour is enriched with boron and the lineal energy distribution measured with the boron-doped TEPC will be similar to that experienced by the tumour cells (Figure 1).

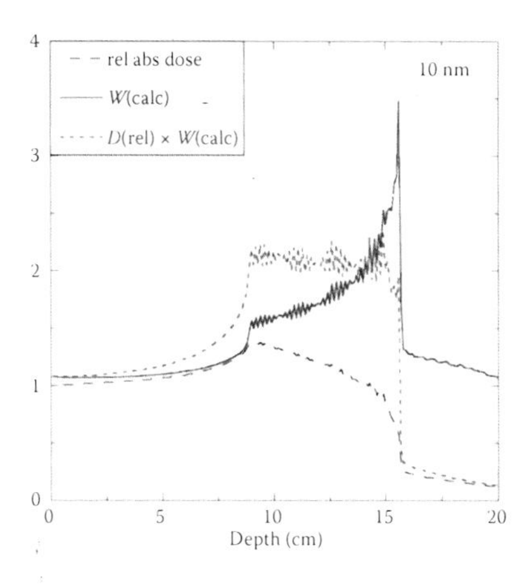

FIGURE 4.19 Depth profiles in water of a 6 cm wide ^{12}C ion beam, 290 MeV u^{-1}. The long dashed line is the absorbed dose profile; the solid line is the calculated weighting factor, W_{calc} (Equation 4.8) with $\bar{y}_D$ values determined for a cylinder of 10 nm height and diameter. The short dashed line is the product of the two other curves and corresponds to the expected biological efficiency in this model. This curve is fairly constant over the SOBP region.

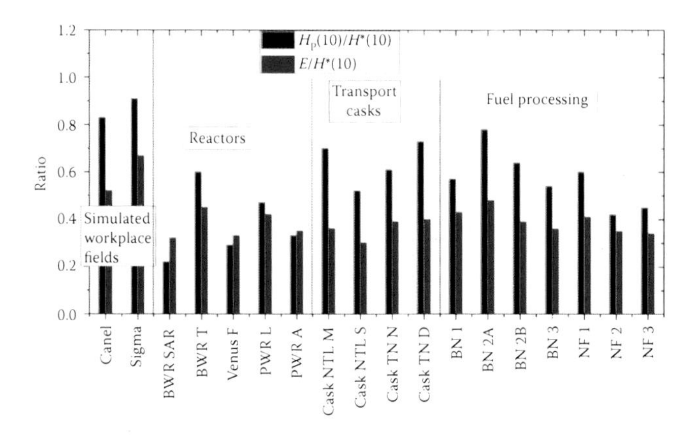

FIGURE 5.3 Ratios of $E/H^*(10)$ and $H_P(10)/H^*(10)$ for the neutron component in mixed fields at a number of locations in the nuclear industry. Also, results for two simulated workplace fields CANEL and SIGMA at IRSN, Cadarache, France, are shown. The reference values were derived from measurements of the neutron energy fluence (Bonner sphere measurements) and its directional distribution [measurements with silicon detectors on a specially designed phantom (Luszik-Bhadra et al., 2004)]. Different positions at a BWR, Krümmel, Germany and a pressure water reactor (PWR), Ringhals Väröbacka, Sweden, as well as transport casks used at the two nuclear power plants, Cask NTL and Cask TN, were investigated. A research reactor, Venus, SCK CEN, Mol Belgium, positions at a fuel processing plant, BN, Belgonucléaire, Dessel, Belgium, as well as at another nuclear facility, NF, were also analysed. The effective dose was derived using the radiation weighting factors, w_R, suggested in ICRP Publication 107 (ICRP, 2007), while tissue weighting factors, w_T, and the specific organs were those defined by ICRP in 1997.

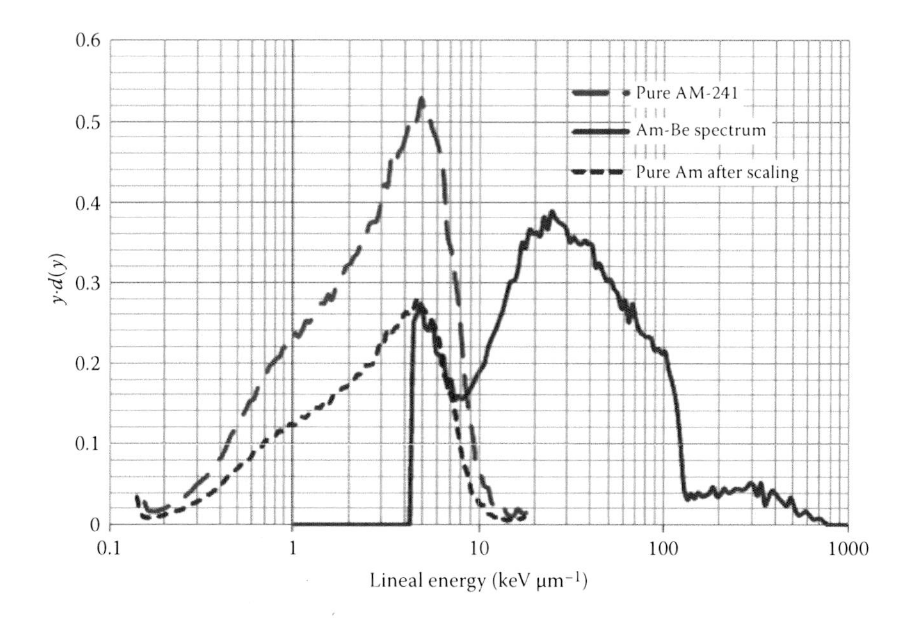

FIGURE 5.7 Method of fitting a pure gamma event-size spectrum to the lower end of a mixed field spectrum in order to extrapolate the measured mixed field spectrum to lower event sizes than are usually obtainable with TEPCs typically used for radiation protection. Once this fitting has been achieved and the extrapolated mixed field spectrum re-normalised the pure gamma spectrum can once again be scaled to fit the lower end of the mixed field spectrum and the scaling factor provides the gamma fraction of the total absorbed dose. Conversely, by subtracting the scaled pure gamma spectrum from the mixed field spectrum one obtains the pure neutron spectrum and the neutron component of the total absorbed dose (Al-Bayati, 2012).

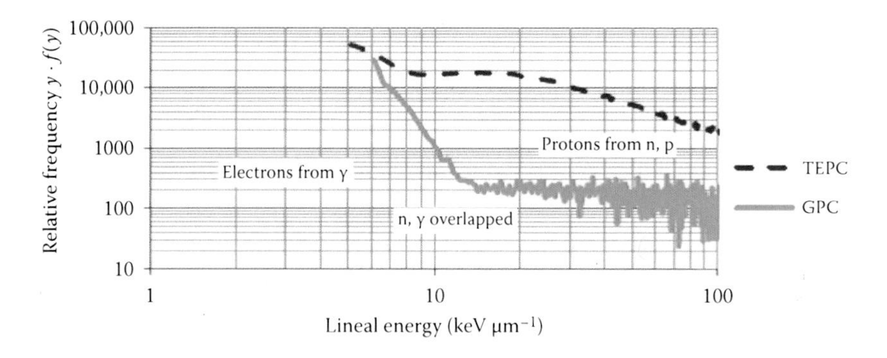

FIGURE 5.8 Overlap region between a TEPC and a graphite-walled proportional counter in the mixed field produced by an Am–Be neutron source. The figure illustrates the much higher degree of separation between electron events and proton events than can be achieved with a TEPC alone (Al-Bayati, 2012).

CHAPTER 5

Applications in Radiation Protection

5.1 INTRODUCTION

In this chapter, the main quantities used in the radiation protection field are reviewed. The protection quantities of major concern in regulation are dealt with first, including a summary of the basic assumptions behind the important quantity effective dose. This quantity is not measurable and to verify dose limits operational quantities have been defined. The operational quantities are related to the protection quantities, and this will be discussed and illustrated. Both the protection quantities and the operational quantities include a weighting factor for radiation quality. For the operational quantities this factor is a function of the linear energy transfer, L. The principal advantage with tissue-equivalent proportional counters (TEPCs) is their ability to measure not only the absorbed dose to tissue but also the lineal energy, y, related to LET and as a consequence the dose equivalent can be estimated. The accuracy with which the dose equivalent may be determined with TEPCs is summarised. Techniques to separate dose contributions from low- and high-LET radiations are also described. In the final section of this chapter a few examples are given to illustrate how the diagnostic capability of experimental microdosimetry can be used.

5.2 HISTORICAL NOTES

The first injuries caused by ionising radiation were recognised very soon after radiation had been discovered (Clarke and Valentin, 2009; Lindell,

1996, 2004) and simple guidance for protection against ionising radiation was quickly recognised as important and necessary. A major problem, however, was the lack of measurable quantities that could be used for setting limits of exposure. On the initiative of the first International Congress in Radiology, held in London in 1925, the British Committee for X-ray Units was asked to establish a corresponding international committee. Such a committee was established during the next meeting of the International Congress in Radiology, which took place in Stockholm in 1928. Today this committee is known as the International Commission on Radiation Units & Measurements (ICRU). During the London meeting a suitable quantity for measurements based on the capacity to ionise air had been suggested and this quantity was accepted during the meeting in Stockholm. Another important committee was established during the same meeting, namely the International X-ray and Radium Protection Committee, which later became known as the International Commission on Radiological Protection (ICRP).

Early injuries reported were skin reactions. Later, exposure to ionising radiation was discovered to be related to cancer incidence. Today activities involving ionising radiation are regulated so that tissue reactions will not occur provided regulations are followed. The probability for cancer or other stochastic effects as a consequence of radiological work is limited by regulations to give a protection that is comparable to what is accepted in other fields of work. The basic recommendations given by the ICRP are internationally accepted and usually implemented by national regulatory bodies that may also provide more detailed and specific regulations to fit circumstances relevant to their country's particular industrial and radiological activities.

5.3 QUANTITIES IN RADIATION PROTECTION

To limit the risk for stochastic effects, a system of protection and operational quantities has been developed. The most important protection quantities are effective dose and equivalent dose to a tissue or organ (ICRP, 2007). Regulations usually refer to these quantities. As they are not measureable, other quantities are needed for operational radiation protection work and have been defined jointly by the ICRU and the ICRP (ICRU, 1993). The quantities are ambient, directional and personal dose equivalent. The protection and operational quantities are related and conversion coefficients between the two, for various irradiation geometries, have been published (ICRP, 2010, 2012; ICRU, 1998).

The ICRP has defined a quality factor, Q, with which the absorbed dose is weighted in such a way that the product of the absorbed dose, D, and Q, represents equal probability for stochastic effects independent of radiation quality. This product is the dose equivalent, H. The quality factor is defined as a function of LET, L. As we have already seen, L is well approximated by lineal energy, y, in volumes of about one to a few micrometres. Methods based on microdosimetry have therefore been useful in complex radiation environments for the determination of operational quantities. Instruments based on these methods are actually the only instruments for direct determination of the quality factor and the dose equivalent.

5.4 PROTECTION QUANTITIES

5.4.1 Equivalent Dose, H_T, to an Organ or Tissue

If the mean absorbed dose to a specified organ or tissue, T, due to the radiation of type R is $D_{T,R}$ then the equivalent dose in the organ or tissue, H_T, is

$$H_T = \sum_R w_R D_{T,R} \tag{5.1}$$

where w_R is the radiation weighting factor for radiation R. The values of w_R are given in Table 5.1. The unit of H_T is J kg^{-1} and has been given the special name sievert (Sv).

The weighting factors for neutrons of different energies are given by the relationships (ICRP, 2007) shown in Figure 5.1.

TABLE 5.1 Recommended Radiation Weighting Factors, w_R (ICRP, 2007), for Different Radiation Qualities

Radiation Type	w_R
Photons	1
Electrons and muons	1
Protons and charged pions	2
Alpha particles, fission fragments and heavy ions	20
Neutrons	A continuous curve as a function of neutron energy, Figure 5.1 and below

Note: All values refer to the radiation incident on the body.

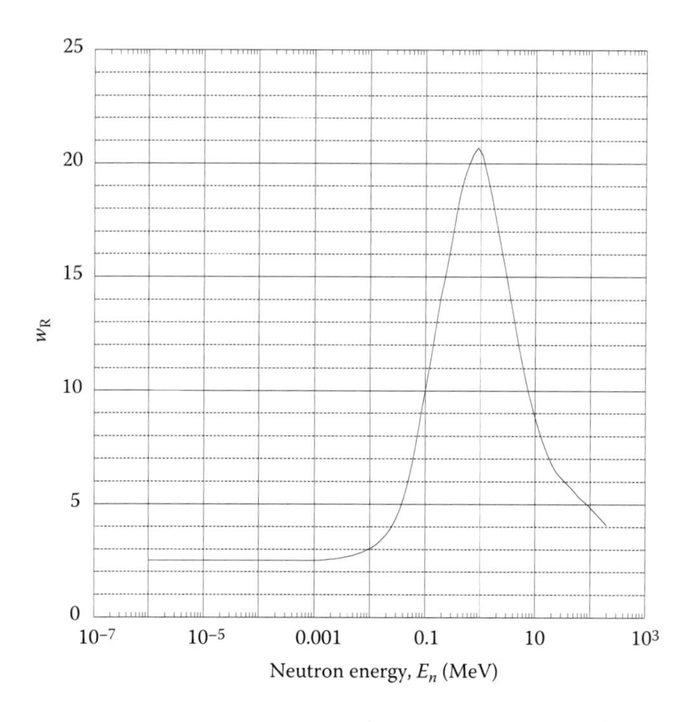

FIGURE 5.1 Radiation weighting factor, w_R, for mono-energetic neutrons (ICRP, 2007).

5.4.2 Effective Dose, *E*

The effective dose is a quantity used for the purpose of setting exposure limits and to implement optimisation for stochastic effects in radiation protection. It is applicable to whole-body irradiation or a partial irradiation for internal as well as external exposures. The effective dose, E, is

$$E = \sum_T w_T \sum_R w_R D_{T,R} = \sum_T w_T H_T \tag{5.2}$$

where w_T is the tissue weighting factor for tissue T and represents the relative detriment* for tissue T and $\sum_T w_T = 1$ (Table 5.2). The sum is performed over all organs and tissues of the human body considered in the definition of the effective dose and for which w_T values are given. The unit for E is J kg^{-1}, which has been given the special name sievert (Sv).

* Detriment is defined by the ICRP as the total harm to health experienced by an exposed group and its descendants as a result of the group's exposure to a radiation source. Detriment is a multi-dimensional concept. Its principal components are the stochastic quantities: probability of attributable fatal cancer, weighted probability of attributable non-fatal cancer, weighted probability of severe heritable effects and length of life lost if the harm occurs (ICRP, 2007).

TABLE 5.2 Tissue Weighting Factors, w_T

Organ/Tissue	Number of Tissues	w_T	Total Contribution
Lung, stomach, colon, bone marrow, breast, remainder	6	00.12	0.72
Gonads	1	00.08	0.08
Thyroid, oesophagus, bladder, liver	4	00.04	0.16
Bone surface, skin, brain, salivary glands	4	00.01	0.04
Σw_T			1.00

Source: ICRP (International Commission on Radiological Protection). The 2007 recommendations of the International Commission on Radiological Protection. Publication 103. *Annals of the ICRP* 37(2–4) (2007).

The concept effective dose was first presented by Jacobi (1975), whose publication is recommended for readers wishing to gain an understanding of this complex quantity. Jacobi's derivation starts with a number of assumptions, which in the lack of precise knowledge have to be used in such a way that the effective dose will most often be an overestimate of the detriment. In short, it is assumed that (1) organs and tissues are not interacting; (2) the probability a stochastic effect will appear after exposure to a specific absorbed dose is independent of the age at which the irradiation occurred; and (3) the absorbed dose is small enough that a linear relation can be assumed between probability of a stochastic effect and the absorbed dose. When these conditions are fulfilled and for radiations with a radiation weighting factor of 1 the total detriment is proportional to E, and

$$E = \sum_{T} w_T D_T \tag{5.3}$$

where D_T, and w_T have the same meaning as earlier. If the total body, TB, with N organs becomes homogeneously irradiated in such a way that the absorbed dose is the same in all organs the effective dose is

$$E = D_{TB} \sum_{T=1}^{N} w_T = D_{TB} \tag{5.4}$$

In a situation of partial irradiation, where only n organs are irradiated, the effective dose is

$$E(n) = \sum_{T=1}^{n} w_T D_T \tag{5.5}$$

When in both whole-body and partial irradiation situations the detriments are the same, i.e., $E = E(n)$, it follows that

$$E = \sum_{T=1}^{n} w_T D_T = D_{TB} \tag{5.6}$$

Thus the effective dose, E, is the whole-body dose that is expected to create the same detriment as is caused by the partial irradiation. The w_T is estimated mainly from the late effects observed among atomic bomb survivors in Hiroshima and Nagasaki.

The effective dose is proportional to the expected stochastic effects in large human populations through probability coefficients (Table 5.3). Although it is not meaningful to apply effective dose to an estimate of the health risk for an individual, effective dose, or more precisely the collective effective dose, is often used when a society wishes to estimate the impact of stochastic effects for a particular radiological event or to compare and optimise different radiation protection practises.

To illustrate the relation between nominal risk coefficient for cancer for the whole population and the effective dose we can estimate the radiological impact of a nuclear power accident where the fall-out is expected to give a population of 1 million people an annual effective dose of 0.2 mSv during a 10-year period as 10^6 (persons) $\times 0.2 \times 10^{-3}$ (Sv year) $\times$ 10 (year) $\times 5.5 \times 10^{-2}$ (Sv – 1) = 110 cancer cases.

According to NCRP Report 121 (NCRP, 1995), where a value of 15 years of loss of life is associated with each fatal cancer, the total impact on the exposed population in our example would result in a maximum of 1650 years of lost life. Owing to the very large uncertainty in the risk coefficients and to the many approximations involved in the definition of the effective dose, this way of evaluating the total health impact for the exposed population is not endorsed by the ICRP (ICRP, 2007), which specifically warns against using collective effective dose in calculating cancer deaths involving trivial exposures to a large population. However, there is some ambiguity in the ICRP recommendations concerning the use of collective effective dose, as their discussion also includes the statement from NCRP Report 121 that

TABLE 5.3 Probability Coefficients for Stochastic Effects (10^{-2} Sv^{-1})

	Total Risk	Induction of Cancer	Genetic Effects
Whole population	5.7	5.5	0.2
Adults	4.2	4.1	

Source: ICRP (International Commission on Radiological Protection). The 2007 recommendations of the International Commission on Radiological Protection. Publication 103. *Annals of the ICRP* 37(2–4) (2007).

when the product of the number of exposed persons and the effective dose is smaller than the reciprocal of the risk detriment, the risk assessment should note the most likely number of excess health effects is zero. In our example, this product is 2000, which is larger than $\frac{1}{5.5 \times 10^{-2}} \approx 20$ and we may conclude that an excess health risk cannot be entirely excluded.

5.4.3 Mean Absorbed Dose

The mean absorbed dose, $\bar{D}_T$, in an organ or tissue, T, is

$$\bar{D}_T = \frac{\int_T D(x,y,z)\varrho(x,y,z)dV}{\int_T \varrho(x,y,z)dV} \quad (5.7)$$

$D(x, y, z)$ is the absorbed dose at the point defined by the coordinates x, y, z in the organ of volume V and density ρ. $\bar{D}_T$ is supposed to be a sufficient measure of the absorbed dose for applications in radiation protection and for estimates of the stochastic effects. Those effects are assumed to have a linear-non-threshold dose–response relationship (LNT model). Usually D_T is used instead of $\bar{D}_T$ to simplify the written form as illustrated in the preceding definitions. Sometimes the dose may be very inhomogeneously distributed in an organ and the mean absorbed dose may be misleading. Several such situations are identified and discussed by the ICRP (ICRP, 2007, Appendix B).

5.5 THE OPERATIONAL QUANTITIES

5.5.1 Dose Equivalent, *H*

As already mentioned the ICRP has defined a quality factor, Q, with which the absorbed dose is weighted in such a way that the product of the absorbed dose, D, to a point in tissue and Q, represents equal probability for stochastic effects independent of radiation quality. This product is the dose equivalent, H, and

$$H = Q(L)D \quad (5.8)$$

Q depends on the linear energy transfer, L, in water and this relationship has been decided by the ICRP and is given in Table 5.4. The unit for H is the sievert, Sv.

In any given situation the absorbed dose will depend on the particulars of the radiation geometry, such as, is the dose determined free in air or is scattering material surrounding the point of measurement? This ambiguity

TABLE 5.4 Quality Factor $Q(L)$

L	Q
<10 keV μm^{-1}	$Q = 1$
10 keV μm^{-1} ≤ L ≤ 100 keV μm^{-1}	$Q = 0.32L - 2.2$
100 keV μm^{-1} < L	$Q = 300/L^{1/2}$

Source: ICRP (International Commission on Radiological Protection). The 2007 recommendations of the International Commission on Radiological Protection. Publication 103. *Annals of the ICRP* 37(2–4) (2007). L_∞ is in water.

had to be overcome by the definitions of the operational quantities that a task group appointed by ICRU and ICRP in the late 1980s was commissioned to establish. Also, at a given location in a radiation field values of the new quantities had to be numerically larger than the values of the effective dose. The last criterion means that if in a particular situation an estimate of the effective dose is looked for, the measured value observed with a survey instrument does not underestimate the effective dose value.

The quantities have been defined by the ICRU (ICRU, 1993). Conversion coefficients between the operational and the protection quantities have been published (ICRU, 1998). In the definitions given in Sections 5.2 and 5.3 for the operational dose quantities, ambient and directional dose equivalent, the terms 'expanded' and 'aligned' radiation fields are introduced along with a phantom called the ICRU sphere. This phantom is introduced to define a geometry that will ensure both sufficient build-up and scattering similar to what would happen in the human body. The sphere has a diameter of 30 cm and the density 1000 kg m^{-3} and has the following constituents in weight per cent: hydrogen 10.1%, nitrogen 2.6%, carbon 11.1% and oxygen 76.2% and is approximately tissue equivalent. The sphere is always centred at the point where the dose equivalent is to be determined. The meaning of the terms 'expanded' and 'aligned' has been explained in ICRP annex B.4 (2007) and can be summarised as follows.

The expansion of the radiation field ensures that the whole ICRU sphere is exposed to a radiation field with the same fluence, energy distribution and directional distribution as the point of interest in the real field. In a hypothetical field which is both expanded and aligned the ICRU sphere is homogeneously irradiated from one direction and the fluence of the field is the integral of the angular differential fluence at the point of interest in the real radiation field over all directions. These conditions and requirements are illustrated in Figure 5.2.

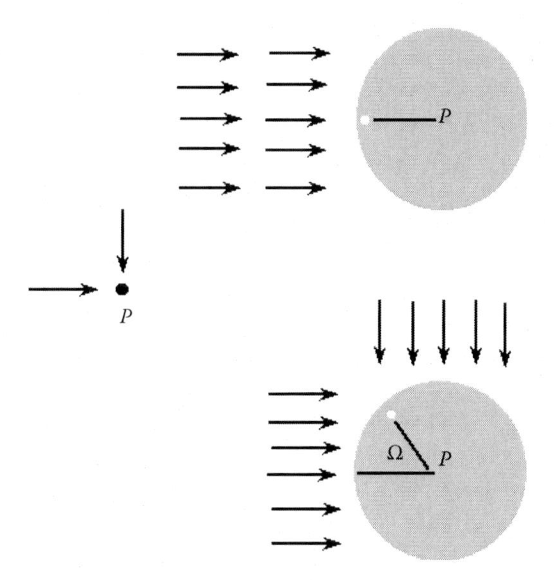

FIGURE 5.2 Radiation field at a point marked by P is to be characterised by the operational quantities ambient dose equivalent, $H^*(d)$ and directional dose equivalent, $H'(d)$. A hypothetical radiation field is created, which in the case of the directional dose equivalent is expanded and in case of ambient dose equivalent is both expanded and aligned. In both definitions the ICRU sphere is centered at the point marked by P and the dose equivalent is determined at a point at depth d, usually 10 mm, indicated by the white spot in the sphere. The position of this point is, in case of $H'(d)$, on a radius that creates an angle Ω to a reference direction, both defined by the user. In case of $H^*(d)$, the position of the point is always on a radius opposing the direction of the parallell radiation field.

5.5.2 Ambient Dose Equivalent, $H^*(10)$

$H^*(10)$, at a point P in a radiation field is the dose equivalent that would have been created by a corresponding expanded and aligned radiation field at 10 mm depth in the ICRU sphere. The 10-mm depth position shall be on a radius of the sphere that is directed towards the radiation field. The sphere shall be centred on the point P. The unit of ambient dose equivalent, $H^*(10)$ is J kg^{-1}, and has also been given the special name sievert (Sv).

The definition of $H^*(10)$ requires that all radiation is coming from one direction as an aligned beam and is expanded so that full scattering from the sphere is achieved (Figure 5.2). However, an instrument with an isotropic response will be independent of the direction of the radiation field and will measure in a multi-directional radiation field as if it were a parallel field. Isotropic instrument response is therefore an instrument requirement for measurements of $H^*(10)$. In practice, radiation fields are often wide enough to cover the whole instrument, which quite often gives sufficient scattering and the need for an expanded field is then satisfied. If the detector has a wall thickness of a few millimetres, the detector will simulate a depth of approximately 10 mm. In practice, therefore, most survey instruments will manage to measure something close to $H^*(10)$. However, no survey instrument measures $H^*(10)$ as it is defined and therefore calibration in a radiation field where the ambient dose equivalent is known is necessary.

5.5.3 Directional Dose Equivalent, $H'(d,\Omega)$

$H'(d,\Omega)$, at a point P is the dose equivalent that would have been created by an expanded field, in the ICRU sphere at the depth, d, on a radius with a special direction, Ω. The unit is J kg^{-1}, and has the special name sievert (Sv).

A depth of 10 mm is a standard depth when penetrating radiation is present, while when the skin is of major concern a depth of 0.07 mm is specified. As before, the sphere is centred at position P and the radiation field is expanded to generate scattered radiation from the whole sphere. The dose equivalent is determined at the depth d in a direction characterised by the angle Ω relative to a user-defined reference direction. It should be intuitively clear from Figure 5.2, that if Ω is allowed to go from 0° to 360°, $H'(d,\Omega)$ will vary with angle a great deal for most radiations. An argument for this quantity is the need to estimate the dose to organs or tissues close to the surface such as the skin or lens of the eye. Originally this quantity was also thought of as quantity that approximates the personal dose equivalent in a calibration situation. In a parallel beam and when $\Omega = 0°$ $H^*(10) = H'(10, 0°)$. In all other irradiation geometries $H^*(10) > H'(10, \Omega)$. Prototypes of instruments aimed at measuring the directional dose equivalent have been tested (d'Errico et al., 2004; Luszik-Bhadra et al., 2004).

5.5.4 Personal Dose Equivalent $H_P(d)$

$H_P(d)$ is the dose equivalent in soft tissue, at depth, d, under a specific point on the body of the person wearing the dosimeter. The unit is J kg^{-1}, and the unit has the special name sievert (Sv).

This quantity is different in the sense that the individual wearing the dosemeter defines its geometry. However, during calibration the dosemeter is positioned on a specified phantom, and calibrated for $H_P(10)$ on the assumption that the body can be approximated by this phantom. For practical reasons calibration of personal dosemeters are usually performed on a slab phantom 30 cm × 30 cm × 15 cm made of Perspex walls and filled with water. In many applications a solid Perspex phantom will do as well (ISO, 1999). It is important to realise that if a person is working with most of the radiation coming from behind and wears the dosemeter on his or her front a severe underestimate of the effective dose may occur. In situations in which the irradiation conditions may be unclear, several dosemeters should be used and positioned around the body (Drake et al., 1998).

5.6 RELATIONSHIPS BETWEEN PROTECTION AND OPERATIONAL QUANTITIES

One intention behind the use of operational quantities is that they should not underestimate the effective dose, E. This can be judged from the Publication 116 by the ICRP (ICRP, 2010). For photon energies above about 3 MeV $H^*(10)$ starts to underestimate E independent of the irradiation geometry. For neutron energies the situation is more complex. There is an energy interval between 3 MeV and 12 MeV where $H^*(10)$ will underestimate E for frontal irradiation. Above 50 MeV there is underestimation of E for irradiations from both the front and back. For neutrons above 75 MeV there is an underestimation of E independent of the irradiation geometry. Neutron fields encountered in radiation protection situations are usually broad spectra and in practice it is seldom a problem that operational quantities are not conservative with respect to the effective dose in a narrow neutron energy interval. An example is given in the text that follows.

In an investigation of radiation fields in the nuclear industry both the ambient and personal dose equivalent were measured. The effective dose was also evaluated (Schuhmacher, 2006). The ratios of $E/H^*(10)$ and $H_P(10)/H^*(10)$ for the neutron component of these mixed fields at different locations are shown in Figure 5.3. In general the ratio E/H^* is less than 0.5 meaning that $H^*(10)$ overestimate E by a factor of 2. The ratio H_P/H^* is at most positions larger than E/H^*, meaning that the values of H_P are closer to E than values of H^*. This is not surprising, as in the definition of H_P not just the depth in the sphere but also the angular distribution of the radiation is taken into account. The following relation can be inferred from the figure: $H^*(10) > H_P(10) > E$. However, there are three exceptions where actually $H_P(10) < E$, but also $H^* < E$. One of the exceptions is the field at position BWR SAR. This position was below the core of the boiling water reactor (BWR) power plant and most radiation was coming from above. At the other positions, Venus F and PWR A, $H_P(10)$ is numerically just a few

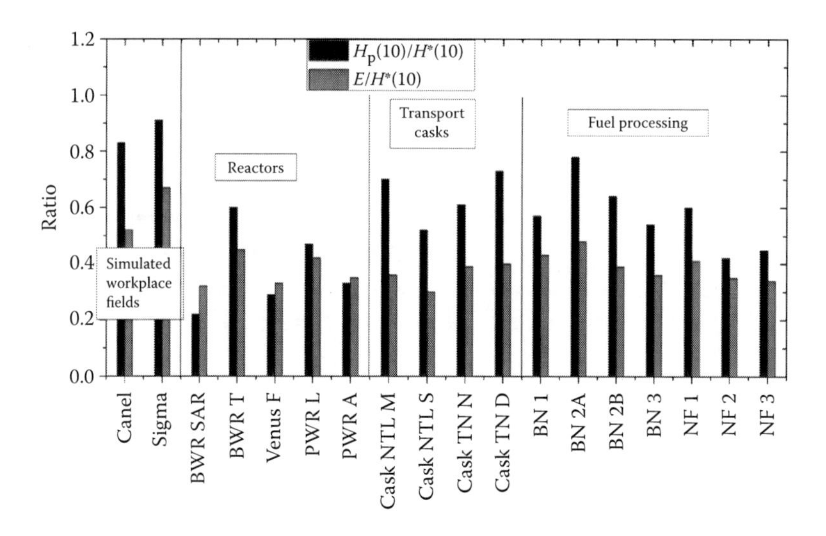

FIGURE 5.3 **(See colour insert.)** Ratios of $E/H^*(10)$ and $H_P(10)/H^*(10)$ for the neutron component in mixed fields at a number of locations in the nuclear industry. Also, results for two simulated workplace fields CANEL and SIGMA at IRSN, Cadarache, France, are shown. The reference values were derived from measurements of the neutron energy fluence (Bonner sphere measurements) and its directional distribution [measurements with silicon detectors on a specially designed phantom (Luszik-Bhadra et al., 2004)]. Different positions at a BWR, Krümmel, Germany, and a pressure water reactor (PWR), Ringhals Väröbacka, Sweden, as well as transport casks used at the two nuclear power plants, Cask NTL and Cask TN, were investigated. A research reactor, Venus, SCK CEN, Mol Belgium, positions at a fuel processing plant, BN, Belgonucléaire, Dessel, Belgium, as well as at another nuclear facility, NF, were also analysed. The effective dose was derived using the radiation weighting factors, w_R, suggested in ICRP Publication 107 (ICRP, 2007), while tissue weighting factors, w_T, and the specific organs were those defined by ICRP in 1997. [Reproduced from Reginatto M., Tanner R. and Vanhavere F. *Evaluation of individual dosimetry in mixed neutron and photon radiation fields*. PTB Berichte PTB-N-49 (2006), with permission from PTB].

per cent below E. We may conclude that a good personal dosemeter is able to estimate the effective dose better than a survey meter. However, as mentioned previously, severe underestimates may occur in particular irradiation situations.

One of the most occupationally exposed personnel categories is found among air crew on-board commercial aircraft. Here the inevitable cosmic radiation can cause an annual effective dose up to a few millisieverts. The radiation environment has been studied in great detail and results have been collected and reported, for instance, in EC (2004) and NASA

(2003). The radiation field is quite predictable and the effective dose can be calculated with different codes (ICRU, 2010). Personal dosemeters are therefore, in general, not needed. Survey meters calibrated for determination of the $H^*(10)$ on board aircraft are used to check the calculations. The EC report concluded that the total uncertainty in calculated values of E and measured $H^*(10)$ were about 30% and 25% respectively (coverage factor 2) (EC, 2004). $H^*(10)$ measured with TEPC's approximates E within these uncertainties.

5.7 DOSE EQUIVALENT AND MICRODOSIMETRY

The effective dose is defined on condition that the absorbed dose is so small that any increment of the absorbed dose has a constant probability per dose unit to cause a stochastic effect (Jacobi, 1975), implying there is a linear non-threshold relation between the detriment and effective dose. It is tempting to associate this linearity to the dose range where the probability for single events is dominating. Here an increase in dose results only in an increase of the number of volumes experiencing one event, while the mean specific energy remains the same. When all volumes have been hit once, a further increase of the dose will lead to a linear increase of the mean number of events per volume. Figures 5.4 and 5.5 show the relation between mean specific energy in volumes experiencing at least one energy deposition event, $\bar{z}_h$, and absorbed dose for spherical volumes with 8 μm diameter for ^{60}Co γ and 14 MeV neutrons. The relative standard deviation in $\bar{z}_h$ is also shown. An 8 μm diameter volume was chosen as ICRP assumes that stochastic effects are initiated at the cellular level and the cell nucleus typically has a diameter of 8 μm. The dose interval where single events dominate is below about 1 mGy in the low-LET beam and below 10 mGy in the neutron beam. The linear non-threshold dose response relation is applied to occupational annual dose values at least up to 50 mSv. For low-LET radiation this corresponds to 50 mGy, a dose value larger than 1 mGy, the dose below which single events dominate for low-LET beams. For high-LET radiation with weighting factors above 5, the 50 mSv limit corresponds to dose values of 10 mGy or less, a dose range at which single events do dominate the energy deposition process in cellular volumes of 8 μm diameter. The limits 1 mGy and 10 mGy will of course change if the critical target size is changed. According to the ICRP (2007), radiation injury repair during irradiation can take place at dose rates below 100 mGy h^{-1}. For low-LET radiation this dose rate roughly corresponds to 100 events per hour in the cell nucleus and some of the cells may eliminate

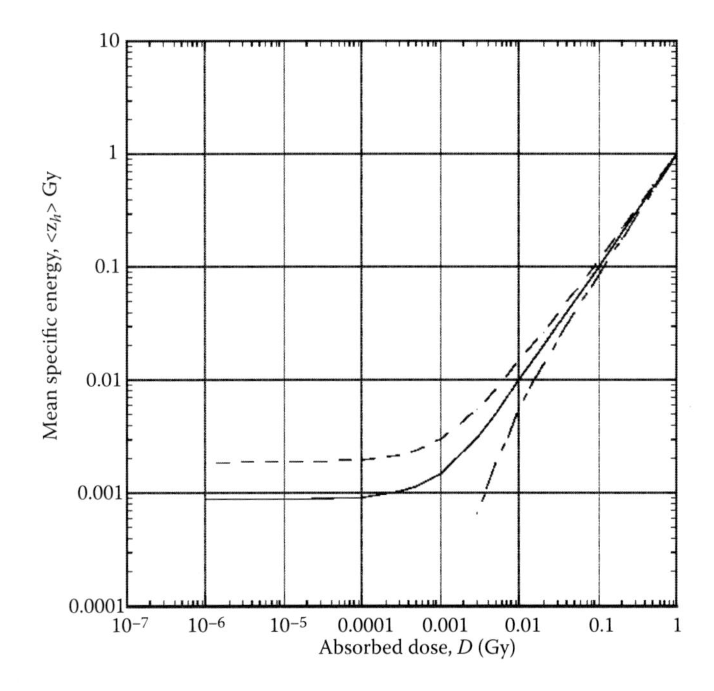

FIGURE 5.4 Mean specific energy, $\bar{z}_h$, in 8 μm volumes which experience at least one energy deposition event for different values of the absorbed dose when irradiated with ^{60}Co γ radiation. The full line is $\bar{z}_h$ and the two other lines show $\bar{z}_h \pm \sigma$ (one standard deviation). The $\bar{z}_h$ is calculated with Equation 2.34 in Chapter 2 with $\bar{z}_s = 0.87$ mGy. The standard deviation is calculated from Equation 2.35 also in Chapter 2 with $\bar{z}_{D,s} = 2.0$ mGy.

the damage caused by these events in this time period. Microdosimetric arguments alone seem not sufficient for predicting a linear non-threshold dose relationship down to low doses but may be useful in exemplifying the outcome of irradiation of biological targets.

We may construct an example to illustrate the relation between absorbed dose and specific energy. Maintenance work inside the containment of a nuclear power plant was planned. The ambient dose equivalent, $H^*(10)$ from gamma radiation of about 1 MeV is observed to be 1 mSv/h; only gammas were of concern. The number of working hours was estimated to be 10 h. The total expected ambient dose equivalent is thus 10 mSv. For low-LET radiation $Q = 1$ and the expected absorbed dose is therefore 10 mGy. With simple tools it was possible to reduce the dose rate a factor 10 and consequently the total dose was reduced to 1 mGy. Two alternatives were then discussed. The work could be split between five workers with each individual receiving 0.2 mGy, or two persons could carry

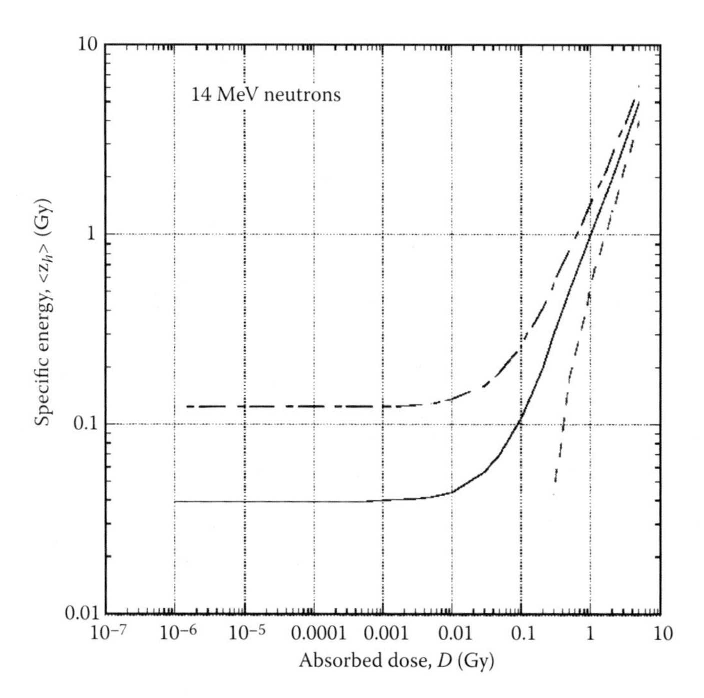

FIGURE 5.5 Mean specific energy, $\bar{z}_h$, in 8-μm volumes experiencing at least one energy deposition event for different values of the absorbed dose when irradiated with 14-MeV neutrons. The full line is $\bar{z}_h$ and the two other lines show $\bar{z}_h \pm \sigma$ (one standard deviation). The $\bar{z}_h$ is calculated with Equation 2.34 in Chapter 2 and $\bar{z}_s$ = 39 mGy. The standard deviation is calculated from Equation 2.35 also in Chapter 2 and $\bar{z}_{D,s}$ = 220 mGy.

out the entire job each receiving 0.5 mGy. Table 5.5 shows the mean specific energy, $\bar{z}_h$, in cell nuclei (8 μm) experiencing an event that correspond to the dose values discussed. At 10 mGy D and $\bar{z}_h$ have the same value, while below about 1 mGy, $\bar{z}_h$ is always numerically larger than D, and approaches $\bar{z}_s$. The absorbed dose, D, to workers in the two situations becomes 0.5 mGy or 0.2 mGy while the mean specific energy to those cell nuclei experiencing an event $\bar{z}_h$, remains almost unchanged (0.9 mGy and 1 mGy respectively). The consequence of the dose reduction (0.5 mGy to 0.2 mGy) is a decrease in number of cell nuclei experiencing an event by a factor of 2.5. As both dose values are well below accepted dose limits, a number of other factors may become decisive for which work plan is chosen. For instance if further work is expected by the persons involved, the number of available persons trained for the work and so on. If the risk for cancer is increasing with dose, the risk is theoretically reduced a factor of 2.5 when the work is split

TABLE 5.5 Absorbed Dose, D, and Mean Specific Energy, $\bar{z}_h$, in Cell Nuclei (8 μm) Experiencing at Least One Single Event When Irradiated in a Beam of ^{60}Co γ-rays

D (mGy)	$\bar{z}_h$ (mGy)
10	10
1	1.5
0.5	1.0
0.2	0.9

between five workers, but as already pointed out the probability for late effects determined from a measure of effective dose and using population-derived risk factors is never applicable to specific individuals.

If the radiation field had been a neutron beam of 14 MeV the graph in Figure 5.5 would have applied. We first need to convert from $H^*(10)$ and Sv to D and Gy. If we assume that $H^*(10)$ is a good approximation for the effective dose, E, a radiation weighting factor can be found from Figure 5.1 and is approximately 7. For this neutron energy $\bar{z}_s$ = 39 mGy (Booz and Feinendegen, 1988). The ambient dose equivalent values in the example above 10, 1 and 0.2 mSv are divided by 7 and the absorbed dose values become 1.4, 0.14 and 0.03 mGy. Figure 5.5 gives the corresponding values of the mean specific energy in those cell nuclei experiencing an energy deposition, which in this situation is the same for all three values and $\bar{z}_h$ = 39 mGy. Thus there is no reduction in the mean specific energy to the cell nuclei although the dose is reduced by a factor of 50. As in the previous example, the number of nuclei hit is instead reduced.

Few radiological workers get more than 2 mSv of low-LET radiation annually and a typical annual occupational dose may reasonably be put at 1 mSv. An annual average dose greater than 20 mSv is above the recommendations from the ICRP. If an annual dose of 6 mSv cannot be excluded the worker is usually equipped with a personal dosemeter. Other dose values of interest that can be used for orientation when discussing radiological risk are the internal annual dose from ^{40}K, which is typically considered to be 0.18 mSv; the annual background dose, often taken to be around 3 mSv; and an acute (instant) dose of 50 mSv for which there is evidence of an elevated cancer risk in humans (Brenner et al., 2003). If the discussion is limited to low-LET radiation (for which Q = 1) dose equivalent (unit mSv) can be exchanged with absorbed dose (unit mGy). The probability that a cell nucleus (8 μm diameter) is hit at least once ($1 - p(0)$) for the absorbed dose values mentioned are listed

TABLE 5.6 Mean Number of Events, $\bar{n}$, and the Fraction of Cells Being Hit $(1 - p(0))$ at Certain Absorbed Dose Values of Interest in Radiation Protection

$$p(\nu) = \frac{\bar{n}^{\nu} e^{-\bar{n}}}{\nu!}$$

	D	0.18 mGy	1 mGy	3 mGy	6 mGy	20 mGy	50 mGy
8 µm	$\bar{n} = D/\bar{z}_s$	0.21	1.15	3.44	6.90	23.0	57
	$1 - p(0)$	0.19	0.68	0.97	1.00	1.00	1.00
1 µm	$\bar{n} = D/\bar{z}_s$	0.002	0.013	0.038	0.075	0.25	0.63
	$1 - p(0)$	0.002	0.013	0.037	0.073	0.22	0.47

Note: The calculation is made for a volume with 8 µm and 1 µm diameter and low-LET radiation (^{60}Co γ).

in Table 5.6. In the same table the mean number of events, $\bar{n}$, is also shown. It is interesting to note that even at 'background' levels (3 mGy), we are already in the regime of every cell nucleus being hit with multiple hits per cell during a year. Of course for high-LET radiation, such as fast neutrons, it is only at doses of 100 mGy and higher that the situation of all cells being hit multiple times arises (Figure 5.5). Results from chromosome aberration studies have in the past influenced the $Q(L)$ relation (ICRP, 2003). A chromosome occupies a volume of about 1 µm in diameter. The corresponding probabilities for this volume are also found in Table 5.6. Here the fraction of volumes experiencing a hit is expected to be about 7% at 6 mGy. This is one way of presenting information on ionising radiation and its interaction in small targets. Discussions of this kind have been published by, for instance, Booz and Feinendegen (1988), Brenner and Sachs (2005), Waker et al. (2015), and Zaider and Rossi (1998).

In the paper by Waker et al. (2015) an analysis was also made for the case of internal exposures of radioactive iodine resulting from a nuclear power plant accident. The calculation was based on having an estimate of the thyroid committed equivalent dose from the exposure and using the internal dosimetry code IMBA (IMBA, 2015) to determine the uptake activity of iodine that would result in that value. From the uptake activity, the total number of beta particle tracks can be derived and using estimates of the range of beta particles of average energy and the size of a thyroid cell the total number of cells hit for all decays delivering the committed equivalent dose can be found. If the number of cells hit is divided by the total number of cells in the thyroid a value for the fraction of cells in the organ experiencing at least one hit is obtained. Reasonable agreement between microdosimetric analysis and conventional internal dosimetry was found in terms of the fraction of cells hit and mean event numbers. This approximate analysis found that even for committed equivalent doses of 0.3 mSv the whole thyroid cell population is affected and each cell experiences on average some 28 energy-depositing events. In this type of microdosimetric analysis dose rate is obviously a crucial factor in trying to assess risk. If we assume an effective half-life of ^{131}I of around 7 days then nearly 90% of the committed dose will have been delivered in 3 half-lives or 504 hours. This means that on average there will be some 18 hours between each event in an individual thyroid cell, which is long compared to common cellular repair mechanisms.

5.8 MEASUREMENT OF OPERATIONAL QUANTITIES USING MICRODOSIMETRIC TECHNIQUES

In radiation protection, measurements are made to ensure that the dose levels are within an acceptable dose range and below limits set by regulation. As described in Section 5.5 the quantity most appropriate for survey instruments is the ambient dose equivalent. A great advantage with a microdosimetric instrument is its capacity to determine this quantity for both low- and high-LET radiations simultaneously. This is because the instruments measure both the absorbed dose and the lineal energy, y, which for volumes in the micrometre range is closely related to LET, L. In most practical situations the radiations of concern are low-LET radiation in combination with neutrons of different energies.

5.8.1 The Pulse Height Analysis Method Applied to Dose Equivalent Measurements

A detailed evaluation of mainly prototype TEPCs in neutron beams was organised by the EURADOS and the Physikalisch-Technische Bundesanstalt (PTB), Braunschweig, in the 1980s (Alberts et al., 1988; Dietze et al., 1988; Menzel et al., 1989). The main objective was to investigate the $H^*(10)$ response of these devices to neutrons. The observations made at that time are still valid and a few of the conclusions are restated here. In the investigation most detectors simulated a tissue-equivalent sphere with diameters in the range of 1 μm to 4.5 μm. Other differences were connected to wall thickness and dimensions. With a thick wall thermalisation of neutrons can lead to capture processes such as the H(n, γ)D reaction, which increases the photon absorbed dose and the

$^{14}N(n, p)^{14}C$ reaction, which add to the absorbed dose at lineal energies below about 140 keV μm^{-1}. For neutrons above 1 MeV neither counter size nor wall thickness affect the spectra or $H^*(10)$ responses significantly, at least up to the maximum neutron energy used in this study of 14.8 MeV. The energy dependence of the quality factor response is quite large below 0.5 MeV and depends to a large extent on the fact that a TEPC measures y not L and the two quantities are different in particular in this energy region due to elastic scattering of protons that have energies too low to cross the detector cavity (Figure 5.6). The large dependence of the H^*-response on neutron energy makes the choice of neutron field for calibration important. Special calibration fields have been developed such as moderated ^{252}Cf fields suitable for measurements in the nuclear

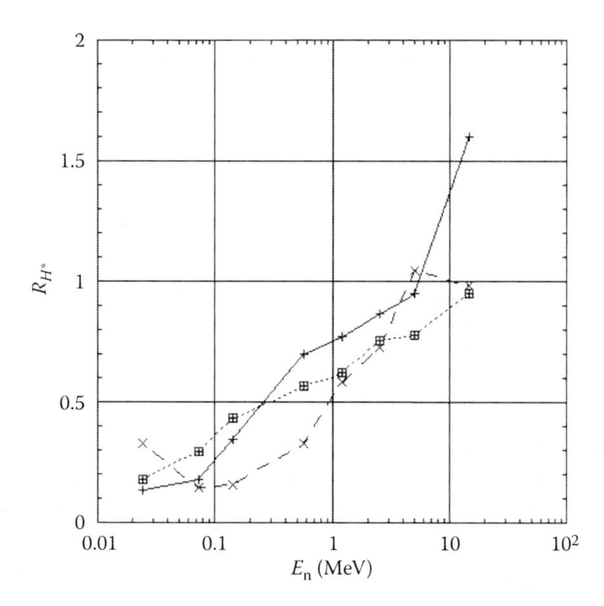

FIGURE 5.6 Ambient dose equivalent response R_{H^*} for three different TEPC systems. The response R_{H^*} is defined as the ratio of measured $H^*(10)$ to the reference value, H^*_{ref}. All three detectors were made of A-150. The results identified with crosses and a long dashed line were measured with a detector furnished with a polyethylene cap of 18 mm placed over the 1-mm thick inner wall of A-150. The response to thermal neutrons (not shown) for the two thin-walled detectors was 0.8 and 0.4 for the thick walled detector. The original data were published by Dietze et al. (1988), utilising a conversion function between $H^*(10)$ and fluence, Φ, published by Wagner et al. (1985). More recent data exist (ICRU, 1998) and the ratio of the old to the new conversion factors between H^* and Φ was used to multiply the original values of R_H to obtain those shown in this figure.

industry and the high-energy field developed at CERN for measurements on board aircraft (Mitaroff and Silari, 2002).

Separation of the low-LET dose from that due to high-LET particles is of clear interest as different quality factors apply. This separation is obscured by a couple of different phenomena. When a charged particle crosses a detector it may cross the gas volume along any of its chord lengths, which means that the energy deposited along a short chord by, for instance, a proton, may be the same as that deposited by an electron crossing the cavity along a longer chord. Another possibility is that the proton from neutron elastic scattering may have an energy (recall that the scattered proton can have any energy from zero up to the maximum energy of the neutron) that gives it a range less than the simulated diameter. As a consequence of both these possibilities the energy deposition of some ion tracks will be low enough for the energy deposition event to appear in that part of the single-event dose distribution that is assumed to be due to electron tracks from photon interaction. The separation of the gamma dose contribution from the rest of the dose distribution therefore needs some care. Two different methods have been typically employed. In the first method the gamma component is determined by 'fitting' a pure gamma dose distribution, typical for the radiation field under investigation, to the lower part of the mixed dose distribution as illustrated in Figure 5.7. The process is usually carried out in two stages. The first stage consists of fitting the appropriate photon spectrum to the lower end of the mixed field spectrum, which for TEPCs typically used in radiation protection is usually measured only down to a few keV μm^{-1}. In this way, a fully extrapolated mixed field spectrum is obtained. If both the extrapolated mixed field spectrum and the pure photon spectrum are normalised so that the area under each spectrum represents unit dose then in the second stage, the photon spectrum is scaled so that it again corresponds to the lower end of the mixed field spectrum and the scaling factor provides the fraction of the total dose due to γ-rays. On subtraction of the scaled photon spectrum the remaining area can be ascribed to the neutron dose. The 'fitting' method does, however, require some prior knowledge of the photon field because the shape of a photon event-size distribution is energy dependent. For the Am–Be neutron field shown in Figure 5.7 the choice of a pure photon spectrum was easily made, as the irradiation arrangement ensured that the 60-keV photons emitted by the ^{241}Am would be by far the dominant photon component of the mixed field. In a real radiation protection scenario the appropriate choice may not be so

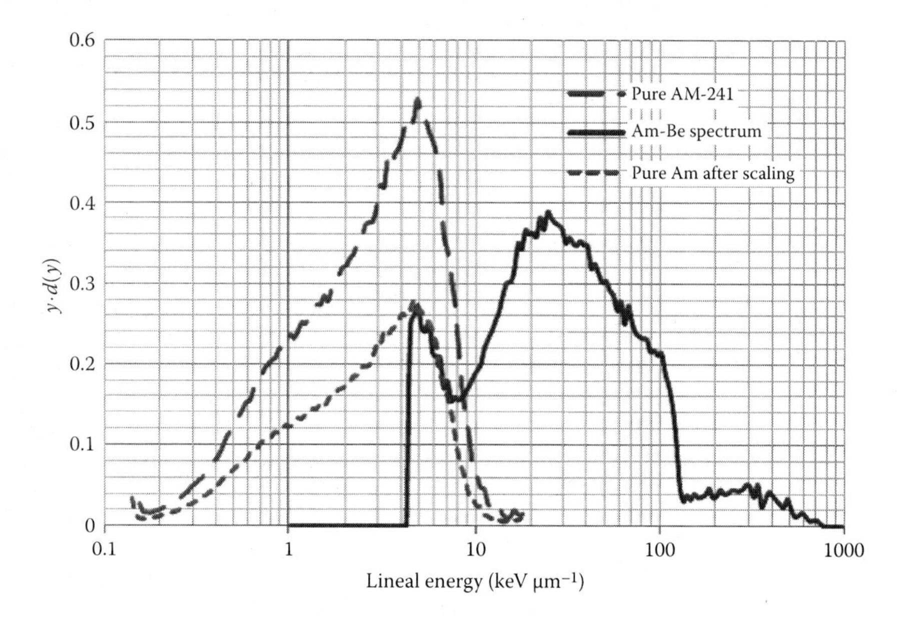

FIGURE 5.7 **(See colour insert.)** Method of fitting a pure gamma event-size spectrum to the lower end of a mixed field spectrum in order to extrapolate the measured mixed field spectrum to lower event sizes than are usually obtainable with TEPCs typically used for radiation protection. Once this fitting has been achieved and the extrapolated mixed field spectrum re-normalised the pure gamma spectrum can once again be scaled to fit the lower end of the mixed field spectrum and the scaling factor provides the gamma fraction of the total absorbed dose. Conversely, by subtracting the scaled pure gamma spectrum from the mixed field spectrum one obtains the pure neutron spectrum and the neutron component of the total absorbed dose (Al-Bayati, 2012).

obvious. To avoid this problem, neutron-insensitive proportional counters such as graphite-walled proportional counters (GPCs) can be used. A comparison between measurements made of an Am-Be source using a TEPC and a GPC is shown in Figure 5.8 (Al-Bayati, 2012). This technique will work for neutron energies below a few MeV (Section 5.8.2).

In the second method of separating the low-LET and high-LET components of a mixed field a lineal energy threshold is set at 10 keV μm^{-1} in the mixed-dose distribution. Anything below the threshold is considered a photon event and anything above a neutron event. For radiation protection purposes the threshold method is likely to be adequate; however, when more accurate gamma-neutron dose determination is required, as in radiobiological experiments, the fitting method will be more appropriate.

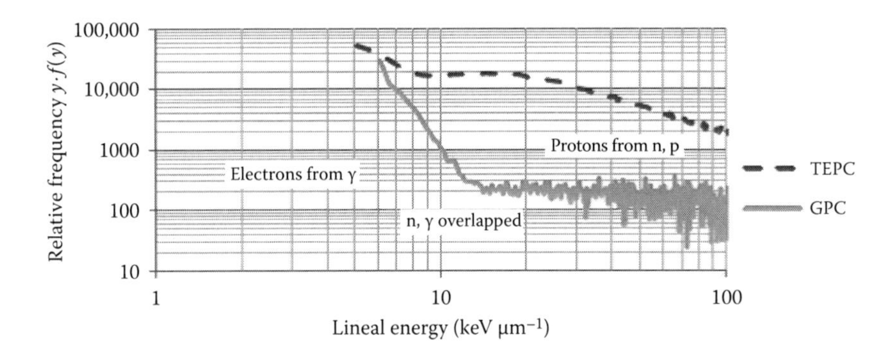

FIGURE 5.8 **(See colour insert.)** Overlap region between a TEPC and a graphite-walled proportional counter in the mixed field produced by an Am–Be neutron source. The figure illustrates the much higher degree of separation between electron events and proton events than can be achieved with a TEPC alone (Al-Bayati, 2012).

In the PHA method of experimental microdosimetry the quality factor is determined from the single-event dose distribution in which the dose in a specific y-interval is given a weight equal to the quality factor in the L-interval indicated in Table 5.4, which is identified by usually setting $y = L$. An average quality factor is then obtained by integrating over the lineal energy dose distribution

$$\bar{Q}(y) = \int_{\min}^{\max} Q(y)d(y)dy \tag{5.9}$$

When this factor is multiplied with the absorbed dose to the detector gas, D, the dose equivalent, H, to the gas cavity is obtained and if $H^*(10)$ is the required quantity then

$$H^*(10) = N_H \bar{Q}(y) D \tag{5.10}$$

where N_H is a calibration factor determined in a calibration laboratory. Usually a volume of one or a few micrometres is simulated. Other algorithms for determining $\bar{Q}(y)$ have also been used in the PHA method such as that based on Equation 5.11 and typically used in the variance method.

5.8.2 The Variance or Variance–Covariance Technique Applied to Dose-Equivalent Measurements

With the variance or variance–covariance (VC) method only the mean values of $\bar{y}_D$ and D become available and no immediate information about different dose components is obtained. The limited proportionality between $\bar{y}_D$ and Q for neutron energies requires some caution. In Section 5.8.2.1 a few suggested relations between the two quantities $\bar{y}_D$ and Q are reviewed.

In principle, the important result of measurements in radiation protection is the total dose equivalent and the need to resolve dose components is of limited practical interest. However, survey instruments usually measure only gamma rays or neutrons, while TEPC-based instruments measure both simultaneously. It is therefore of some interest to resolve such components to facilitate a comparison to other instruments. A few methods for separation of dose components are described in Section 5.8.2.2.

An important consideration when designing a proportional counter for this method is the detector dimensions. The relation between relative variance, V_r, and $\bar{z}_{D,s}$ assumes that the variance is due to multi-events. Usually this is fulfilled if on average at least 5 to 10 events are observed. As an example, an instrument was designed for measurements on board aircraft, where the dose rate is only a few microsieverts per hour. The chosen interior dimensions of the cylindrical detector were 10 cm diameter and 10 cm height. The A-150 wall thickness was 5 mm and a vacuum container with a wall thickness of 2 mm Al housed the detector. The simulated mean chord length was 2 μm. Some results with this detector are reported in the text that follows.

The advantages with the method are its capacity to measure in pulsed beams (Mayer et al., 2004) and the relatively simple and compact technical equipment needed. Actually the first application of the method was suggested to be for radiation protection measurements in mixed radiation fields (Bengtsson, 1970). More recently, Braby used the same argument when he developed an instrument based on this technique for installation on board spacecraft (Braby, 2015). Few laboratories have applied this method to radiation protection measurements and detailed investigations have up to now been reported mainly by researchers at the Swedish Radiation Protection Authority (SSM).

5.8.2.1 Relations between $\bar{y}_D$ and Q

The cylindrical detector mentioned was tested in several mono-energetic neutron beams, a moderated and un-moderated ^{252}Cf beam as well as two quasi mono-energetic beams of higher energies. The dose mean lineal energy in the beams was determined by Kyllönen et al. (2001) and is plotted in Figure 5.9 as a function of the neutron energy. As LET for the

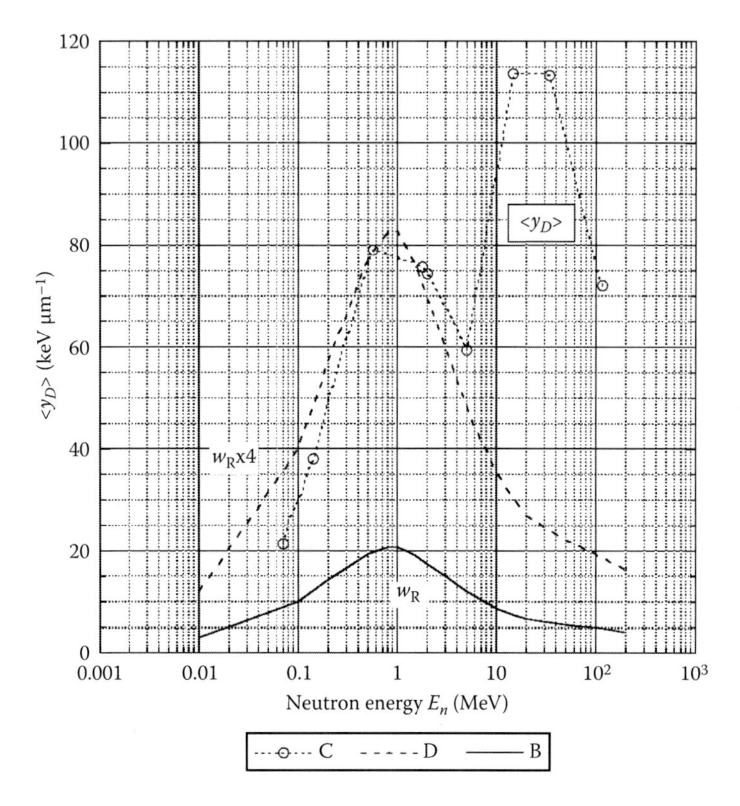

FIGURE 5.9 Measured values of the dose mean lineal energy, $\bar{y}_D$, in different mono-energetic or quasi mono-energetic neutron fields. Also shown is the radiation weighting factor, w_R, for neutrons from Table 5.1. This weighting factor multiplied by 4 is also included in the graph.

beams was unknown, a comparison with w_R is made. This function for neutrons is included in the figure that also includes w_R multiplied by 4. $\bar{y}_D$ seems to approximate the radiation weighting factor after the factor has been multiplied by 4 up to about 5 MeV. At larger neutron energies $\bar{y}_D$ overestimates this function. This relationship may be useful, but seems not to have been used.

Instead a linear relation for determining Q from $\bar{y}_D$ has been applied and is applicable in limited neutron energy ranges with different sets of a and b:

$$Q = a + b\bar{y}_D \tag{5.11}$$

From the experimental results, above pairs of a and b were derived for different neutron energy intervals (Table 5.7). In Figure 5.10 the response to $H^*(10)$ is plotted for the different sets of a and b values as well as a fourth relation called Q_3 comb, which is explained in the text that follows.

TABLE 5.7 Relations Between Q and $\bar{y}_D$ and the Neutron Energy Ranges in Which the Relation Leads to a $H^*(10)$ Response within ± 50%

$Q = a + b\bar{y}_D$	Neutron Energy Range for Which the Response Is Within ± 50%
$Q_1 = 0.52 + 0.28\bar{y}_D$	0.1 MeV–2 MeV
$Q_2 = 0.73 + 0.17\bar{y}_D$	0.2 MeV–6 MeV
Q_3 comb	0.2 MeV–100 MeV
$Q_4 = 0.88 + 0.09\bar{y}_D$	2 MeV–100 MeV

Note: The values were determined for a cylindrical TEPC with the internal diameter and height equal to 10 cm. The detector simulated a mean chord length of 2 μm (Kyllönen et al., 2001; Lillhök, 2007b). The response to a ^{137}Cs gamma source was close to one with all combinations of *a* and *b*.

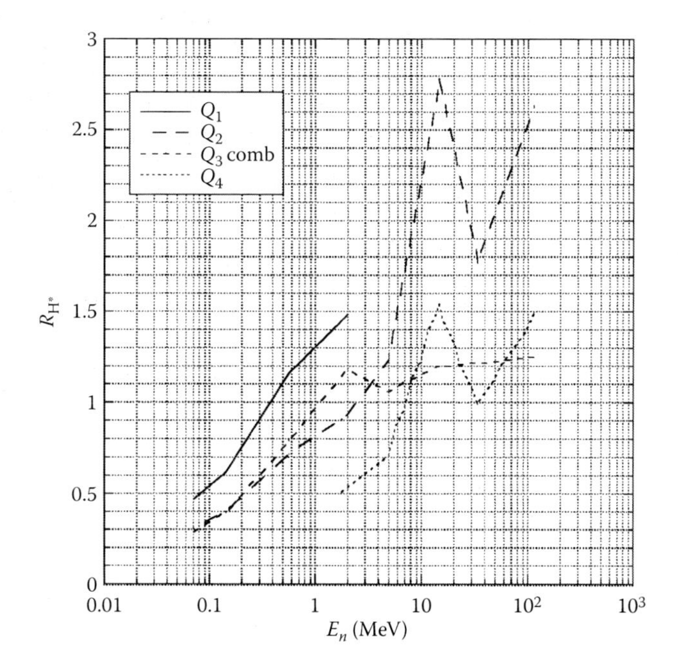

FIGURE 5.10 Measured $H^*(10)$ response to mono-energetic neutron beams when different $Q = a + b\bar{y}_D$ relationships applied. The detector is the same as used for Figure 5.9. The response to thermal neutrons (not shown) was 1.1 (Q_1), 0.7 (Q_2 as well as Q_3 comb) and 0.4 (Q_4) and was calculated with $\bar{y}_D = 41$ keV μm^{-1} (Lillhök, 2007b) in the thermal beam (including the gamma component).

With all pairs of *a* and *b*, $Q = 1$ in a γ ^{137}Cs beam and it has been practical to calibrate the instrument for $H^*(10)$ in a ^{137}Cs beam independent of the measurement application. It is usually possible to roughly estimate the range of neutron energies at a workplace and a suitable pair of *a* and *b*

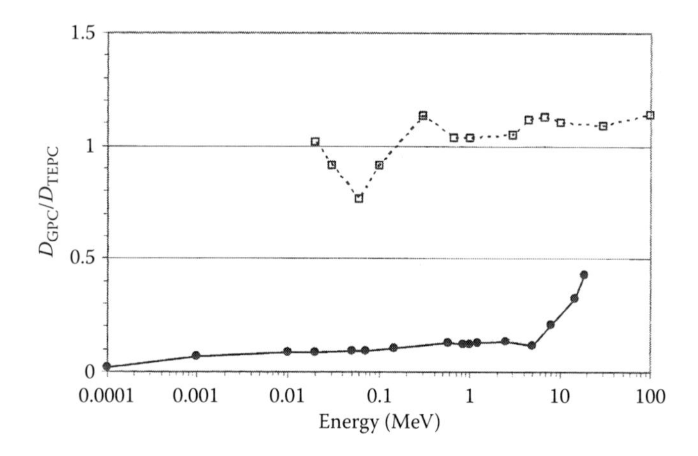

FIGURE 5.11 Calculated values of the ratio between the dose to the graphite detector and the TEPC detector as function of photon energy (open squares) and neutron energy (filled circles) (Kyllönen and Lindborg, 2007).

may be chosen from Figure 5.10 and Table 5.7, and the uncertainty in a determination of $H^*(10)$ from neutrons is estimated to be within ± 50%. Usually the contribution to $H^*(10)$ from neutrons is less than 50% and the uncertainty will in practice be less.

The overestimate of $H^*(10)$ at neutron energies above about 5 MeV seen in Figure 5.10 can be overcome in radiation fields where the number of very large events, n_h, is so small that it is unlikely that not more than one such event will appear in a sampling interval, usually 0.1 s long. These few very large events can be considered as single events added to a multi-event lognormal distribution (Lillhök, 2007b). The y value of this single event is then set equal to L and a Q value is determined from the relation defined in Table 5.4. This possibility, called Q_3 comb in Figure 5.11, is a great improvement as compared to Q_2. A situation when this relation applies is in measurements on board aircraft.

5.8.2.2 Methods to Separate Low- and High-LET Dose Components

A few techniques to separate dose components have been reported (Kyllönen and Lindborg, 2007; Lillhök, 2007b). They are briefly summarised below. Some Geiger–Müller counters have a low sensitivity for neutrons and could in principle be used for determining the gamma dose component; however, experiences from the measurement campaigns are less than convincing. Geiger–Müller counters often over-respond to high-energy photons, which are present in nuclear power environments.

Method 1. Traditionally a neutron dose is separated from the gamma dose component by measuring with two chambers, one of which is ideally insensitive to neutrons. In practice, there is usually a remaining sensitivity to neutrons and to separate the neutron dose component the relationships between the two detector signals are needed. A detector made of pure graphite (Gr) is often used as a complement to the TEPC. The response of the graphite counter relative to that of the TEPC to gamma rays and neutrons are $r_\gamma = \frac{D_{\mathrm{Gr},\gamma}}{D_{\mathrm{TE},\gamma}}$ and $r_n = \frac{D_{\mathrm{Gr},n}}{D_{\mathrm{TE},n}}$, respectively. The absorbed dose to the TE gas in the TEPC has a dose component from gammas, D_γ, and one from neutrons, D_n, and

$$D_{\mathrm{TE}} = D_\gamma + D_n$$

The absorbed dose to the gas in the graphite detector is

$$D_{\mathrm{Gr}} = r_\gamma D_\gamma + r_n D_n \tag{5.13}$$

The neutron absorbed dose to the TE gas, D_n, is derived from both Equation 5.12 and Equation 5.13 and is

$$D_n = D_{\mathrm{TE}} - D_\gamma \tag{5.14}$$

$$D_n = \left(D_{\mathrm{Gr}} - r_\gamma D_\gamma\right)/r_n \tag{5.15}$$

Setting Equation 5.14 equal to Equation 5.15 and solving for D_γ leads to

$$D_\gamma = \frac{D_{\mathrm{Gr}} - r_n D_{\mathrm{TE}}}{r_\gamma - r_n} \tag{5.16}$$

The dose fraction of neutron absorbed dose to the total absorbed dose to the TE gas in the TEPC, d_n, is finally obtained by substitution of D_γ from Equation 5.16 into Equation 5.14, yielding

$$d_n = \frac{D_n}{D_{\mathrm{TE}}} = \frac{r_\gamma - \frac{D_{\mathrm{Gr}}}{D_{\mathrm{TE}}}}{r_\gamma - r_n} \tag{5.17}$$

Thus d_n is derived from the ratio of the two dose values measured with the TEPC and the graphite detector and the dose response ratios r_n and r_γ. Figure 5.11 shows such ratios used for a TEPC filled with propane-based TE-gas and an identical in size carbon-walled proportional counter also filled with propane-based TE gas (Kyllönen and Lindborg,

2007). The figure shows the low sensitivity to neutrons for the graphite detector below a few MeV. For higher energies, the response increases rapidly.

Once d_n is known $\bar{y}_D$ can be split into $\bar{y}_{D,n}$ and $\bar{y}_{D,\gamma}$, if $\bar{y}_{D,\gamma}$ is known. This value is expected to be in the interval 1.5 to 3 keV μm^{-1}. Equation 2.53 in Chapter 2 can be written as

$$\bar{y}_D = (1 - d_n)\bar{y}_{D,\gamma} + d_n\bar{y}_{D,n} \tag{5.18}$$

and

$$\bar{y}_{D,n} = \frac{\bar{y}_D - \bar{y}_{D,\gamma}(1 - d_n)}{d_n} \tag{5.19}$$

With $\bar{y}_{D,n}$ known the dose equivalent can be calculated for the neutron dose component, that is, the appropriate Q_n can be obtained from the $\bar{y}_{D,n}$ value, as discussed in Section 5.8.2.1, and the neutron dose equivalent will be given by

$$H_n = Q_n \times d_n D_{\mathrm{TE}} \tag{5.20}$$

Method 2. When both detectors are tissue equivalent this method may be useful. Here not only $\bar{y}_{D\gamma}$ but also $\bar{y}_{Dn}$ have to be estimated. Since the contribution to $\bar{y}_D$ and D from neutrons decreases rapidly with decreasing neutron energy, the neutron contribution to the total $\bar{y}_D$ decreases even more rapidly with decreasing neutron energy (Equation 2.55). In many neutron fields, $\bar{y}_{Dn}$ is therefore dominated by the fast neutrons, and the expected $\bar{y}_{D,n}$ is then often in the interval 60 to 80 keV μm^{-1} (see Figure 5.9). Equation 5.18 may be written as

$$d_n = \frac{\left(\bar{y}_D - \bar{y}_{D,\gamma}\right)}{\left(\bar{y}_{D,n} - \bar{y}_{D,\gamma}\right)} \tag{5.21}$$

With chosen values for $\bar{y}_{Dn}$ and $\bar{y}_{D\gamma}$ together with the measured $\bar{y}_D$ inserted in Equation 5.21 the dose fraction from neutrons, d_n, is given by this equation. It turns out that d_n is actually not very sensitive to the chosen value of $\bar{y}_{D,n}$, since an error in $\bar{y}_{D,n}$ influences the neutron absorbed dose and the neutron quality factor in opposite directions.

Methods 1 and 2 were applied to results obtained during the measurement campaign within the nuclear industry mentioned earlier (Section 5.6

and Figure 5.3). No significant difference was observed between the two methods (Kyllönen and Lindborg, 2007).

Method 3. This method assumes that the frequency of low-LET events dominates and their frequency distribution can be fitted by a lognormal distribution. If such a distribution is forced to fit the initial slope and the peak in the experimental data, it will in the presence of a high-LET component not fit the experimental data points on the final slope. The difference is due to high-LET events and the integral of this difference is the dose due to the high-LET events.

A test of all three methods was made on board a dedicated flight in which the aircraft circled two geographical points each at two different altitudes (Lillhök et al., 2007c). On the same flight four TEPC instruments measured the single-event distributions and the results were compared with the VC method. Because of the high-energy neutrons present on board aircraft, the TEPC + GPC measurement system failed in resolving the two components. For the two other methods the separation into dose fractions from low- and high-LET was in agreement with what was observed with single-event analysis. The quality factor was also evaluated. On average, the mean quality factor agreed between all instruments within ±15%. This value was reduced to half that value at the higher latitude where the statistics were better because of the increased dose rate. The VC instrument was on average a few percentages below most of the PHA-based instruments.

For aviation dosimetry, the combined standard uncertainty of type A and B in the VC method was reported to be 20% in $H^*(10)$ for a 1-h measurement on-board an aircraft. Uncertainties considered are shown in Table 5.8 (EC, 2004).

TABLE 5.8 Components of One Standard Uncertainty in Measurement of Dose Equivalent with a TEPC based on the VC Method

Influence Factor	**Type A (%)**	**Type B (%)**
Dose	1	
Quality factor (1 h)	6–13	
Temperature dependence		2
Calibration		5
W/*e*		5
Approximation in the method		10
TEPC response characteristics		10
Quadratic sum	6–13	16
Combined A and B	17–21%	

5.8.3 Applications to Radiation Fields at Workplaces

Neutron fields at nuclear power plants have been investigated in detail with reference instruments (Bonner spheres) and different survey instruments such as TEPCs. An early measurement campaign was made at a pressurised water reactor (PWR) at Ringhals, Sweden. The neutron fluence inside its containment was characterised by less than 10% above 100 keV while this fluence component contributed to almost 50% of the $H^*(10)$. Around fuel transport casks the corresponding values were 15% and 75%. The contribution to $H^*(10)$ above 1 MeV was on average 3% and 7% at the two different locations. Figure 5.12 shows single-event dose distributions from one position inside the PWR measured with three different TEPCs. The distributions are dominated by the dose fraction from low-LET radiations (<10 keV μm^{-1}) (Bartlett et al., 1999; Lindborg et al., 1995). When the neutron component to $H^*(10)$ measured by TEPCs was compared to the same results determined with the Bonner spheres systems, the ratio was 0.7 for the PWR positions. The explanation for the difference is the large fraction of the dose equivalent from neutrons below 100 keV, where the TEPC response is low (Figure 5.6). With a different calibration procedure

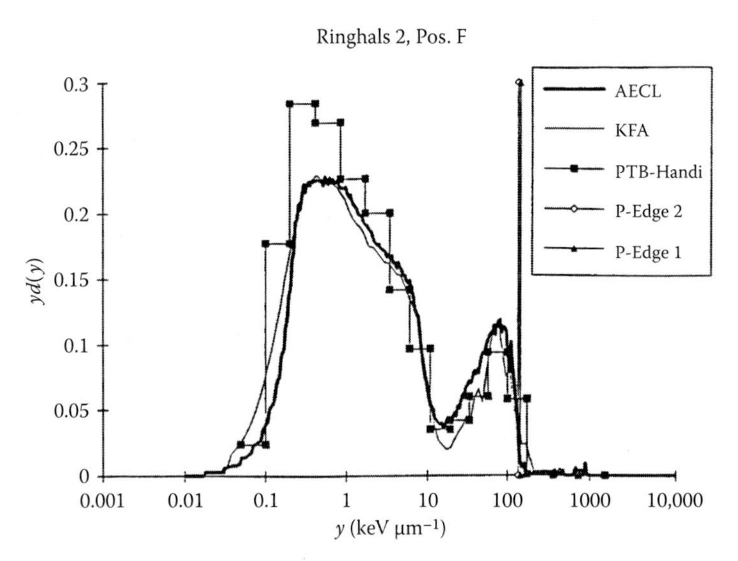

FIGURE 5.12 Single-event dose distributions measured by three different TEPCs at a location inside the containment of a BWR at Ringhals (Bartlett et al., 1999). The three measurements were carried out with TEPC systems from Atomic Energy of Canada Limited (AECL); Kernforschungsanlage - Julich (KFA); and Physikalisch - Technische Bundesanstalt, Braunschweig (PTB). The distributions are dominated by the dose contribution from low-LET radiations (<10 keV).

this could have been overcome. Only the total dose equivalent was evaluated by the VC-based instrument and was on average a few percent below the results of the other TEPCs, but the difference was not judged significant.

In a similar investigation to that reported in Section 5.6, reference values were again determined with a system of Bonner spheres. The only TEPC that took part was the VC instrument described previously consisting of an A-150 detector and a graphite detector (Schuhmacher et al., 2006). The neutron ambient dose equivalent reported by the VC instrument was on average 0.8 ± 0.7 (2 standard deviations) of the reference values. The total ambient dose equivalent was 0.9 ± 0.5 (2 standard deviations) of the reference values. It was concluded that VC measurements of the neutron dose-equivalent component were accurate within the requested uncertainty of ±50% (Lindborg et al., 2007).

Several different TEPC instruments took part in a European Commission contract to map aircrew ambient dose equivalent for many different flight-routes. Figure 5.13 (EC, 2004) shows a single-event dose distribution (dashed line) as well as the dose-equivalent distribution (full line) from these measurements. The contribution to $H^*(10)$ from neutrons and protons is more than 50% on board aircraft and may contribute as much as 80% to the effective dose (E). The neutron fields on board aircraft are in an energy range in which a TEPC operates well.

From the same investigation results of the dose rate and ambient dose-equivalent rate on board aircraft is shown as function of flight time. The flight went from a southern departure point (Bangkok in Thailand) to a northern destination (Copenhagen, Denmark) and was measured with a TEPC (Figure 5.14) (Kyllönen et al., 2001). A stepwise increase in dose occurs as the aircraft gains altitude and a gradual increase occurs when flying farther to the north. These increases are due to the increase of intensity by the cosmic rays. Each point is the average result of a 5 minute long measurement. The large fluctuation in the ambient dose equivalent rate is due to the few high-LET particles depositing energy. The dashed line in Figure 4.13 shows that dose contribution from events above 10 keV μm^{-1} is small and consequently caused by very few events, which confirms the large fluctuation in $H^*(10)$ (Figure 5.14). From Figure 5.14 we may extract that the dose rate increased from 0.72 $\mu Gy\ h^{-1}$ at 28,000 feet and the geographical position N22 E088 to 1.25 $\mu Gy\ h^{-1}$ at 35,000 feet and N30 E015 while Q remained 1.4 at both positions. Towards the end of the tour N56 E016 the altitude was still 35,000 feet but the dose rate was almost

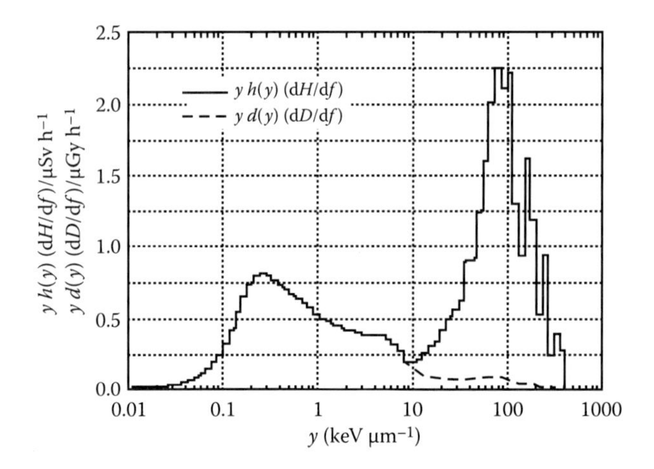

FIGURE 5.13 Single-event dose distribution measured on board aircraft. Above 10 keV μm^{-1} the dose distribution is marked with a dashed line. Thus the low-LET dose component is very large compared to the dose from high-LET events. The full line indicates the dose-equivalent distribution. The high-LET component is now seen to be responsible for 54% of the whole dose equivalent. In this presentation the two distributions are scaled to show $\mu Gy\ h^{-1}$ and $\mu Sv\ h^{-1}$ (Schrewe, in EC, 2004).

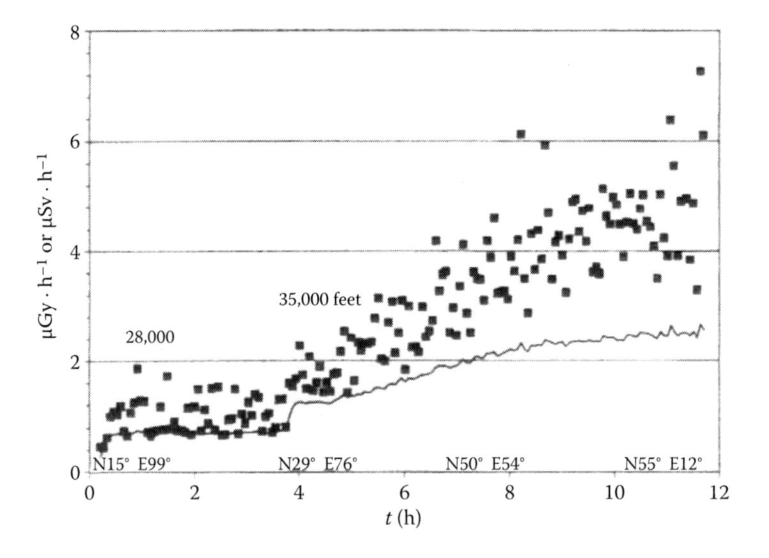

FIGURE 5.14 Results from VC measurements with a TEPC on-board a commercial aircraft between Bangkok and Stockholm on 16 December 1998. The ambient dose equivalent rate (filled square) as well as the absorbed dose rate (solid line) as function of flight time are shown. The flight level is given in feet. The geographic coordinates are shown on top of the x-axis (Kyllönen et al., 2001).

twice as large, 2.64 μGy h^{-1}, and Q had increased to 2. As a consequence, $H^*(10)$ increased from 1 μSv h^{-1} to 5.3 μSv h^{-1} for this flight.

5.8.4 Other Radiation Protection Issues Using Experimental Microdosimetry

The TEPC may be used for more academic purposes when a fundamental radiation question needs to be clarified. For example, before the ICRP 60 recommendations the ICRP had Q = 5 for thermal neutrons. However, when such neutrons penetrate tissue the (n, γ) reaction in hydrogen creates a large dose component of low-LET radiation and the Q = 5 value was questioned as to whether it was the most appropriate value. To illustrate the change of the dose components when low-energy neutrons impinge on tissue, single-event dose distributions were determined with different sleeve thicknesses around a TEPC. The resulting dose distributions are shown in Figure 5.15 (Schuhmacher et al., 1985). These spectra illustrate quite clearly the contributions to the absorbed dose from 2.2 MeV hydrogen capture γ-rays and the 546-keV proton released in nitrogen capture. The large proton dose component observed on the surface of the body decreases at a larger depth due to the attenuation of the thermal neutron fluence and is replaced by an increased gamma-dose component, which has a Q value of 1.

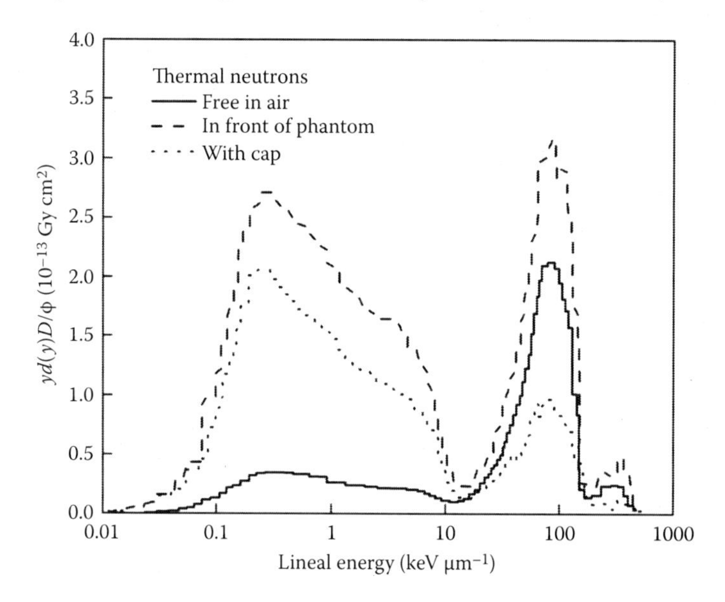

FIGURE 5.15 Single-event dose distributions measured at 2 μm for thermal neutrons measured free in air (solid curve), in front of a phantom (long dashed curve) and with a cap on the detector (dotted line) (Schuhmacher et al., 1985).

The increase in the gamma component comes about because the high-energy capture gamma rays are not strongly attenuated with increasing depth and are able to interact and deposit energy. As a consequence of this increased contribution from low-LET γ-rays the ICRP subsequently adopted a value of $Q = 2.3$ for thermal neutrons.

A further example of the application of experimental microdosimetry addressing fundamental questions in radiation science is the determination of neutron kerma factors, K (kinetic energy released per unit mass of target material by secondary charged particles) for tissue elements such as carbon, oxygen and nitrogen at neutron energies for which they were not well established. The need for this type of information became apparent with increasing interest in the use of high-energy neutrons for therapy and for radiation protection in workplace environments where high-energy neutrons were expected, such as high-energy particle accelerators and high-altitude aviation or space exploration. Schrewe et al. (2000) approached this problem by constructing low-pressure proportional counters with different wall materials, a method pioneered by Scott et al. (1990). By measuring the absorbed dose to the counter gas and converting this value using cavity-theory to absorbed dose in the wall material, coupled with a separate determination of neutron fluence, ϕ, Schrewe and colleagues were able to determine the kerma factors (K/ϕ) for the elements of interest up to neutron energies of 66 MeV. The most experimentally ingenious facet of this work was the determination of K for oxygen and nitrogen. In these cases, pairs of proportional counters were constructed with Al and Al_2O_3, Si and SiO_2 walls as well as Al and AlN and the wall-absorbed dose measurements subtracted to obtain the absorbed dose to the additional element in each case, that is, oxygen for the case of the Al and Al_2O_3, Si and SiO_2 counters and nitrogen for the Al and AlN counters.

Figure 5.16 gives examples of the resulting fractional kerma distributions for oxygen resulting from the subtraction of the dose distributions measured with the matched counters Al and Al_2O_3, Si and SiO_2 from 15-MeV neutrons to 66 MeV.

After the Chernobyl accident, hot particles were found in the environment. To clarify the dose gradient around the particles as well as the radiation quality of the particles an investigation using experimental microdosimetry was made. A hot particle from fallout was fixed on a piece of tape and the dose and single-event distribution investigated with a wall-less TEPC enclosed in a large tank. The single-event dose distribution (Figure 5.17) shows that the radionuclides associated with the particle were emitting

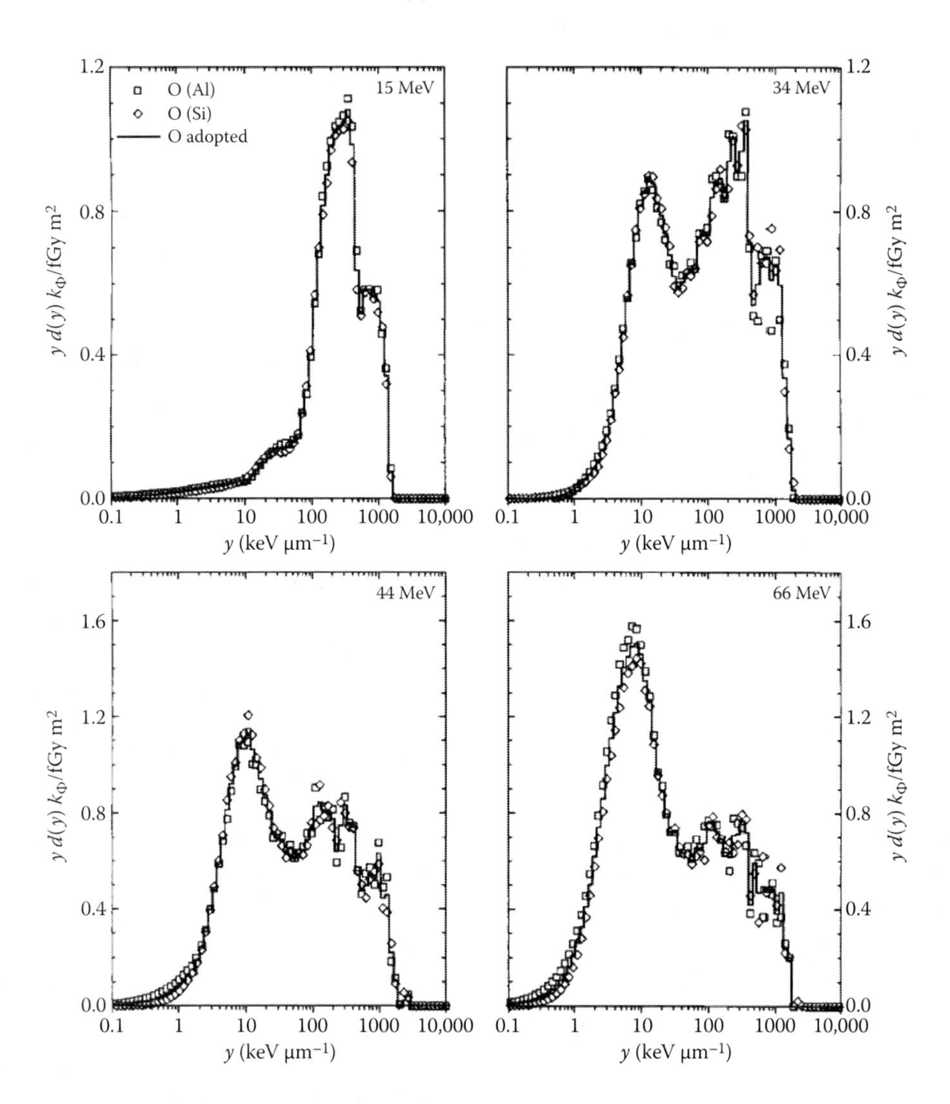

FIGURE 5.16 Dose distributions expressed as kerma fractions $yd(y)K_{\Phi}$ for oxygen derived from measurements with detector pairs of Al and Al_2O_3, Si and SiO_2 (Schrewe et al., 2000).

low-LET radiation. The nuclides were identified by spectroscopic methods and the activities also determined. From this information, the dose rate was calculated and compared with the dose rate observed with the TEPC at distances from 1.4 μm to 15 μm from the source (Figure 5.18) (Grindborg et al., 1990). An agreement within 50% was achieved. The large difference at the shortest distance was claimed to be due to absorption within the source of 10-keV electrons emitted by ^{106}Ru, one of the identified nuclides.

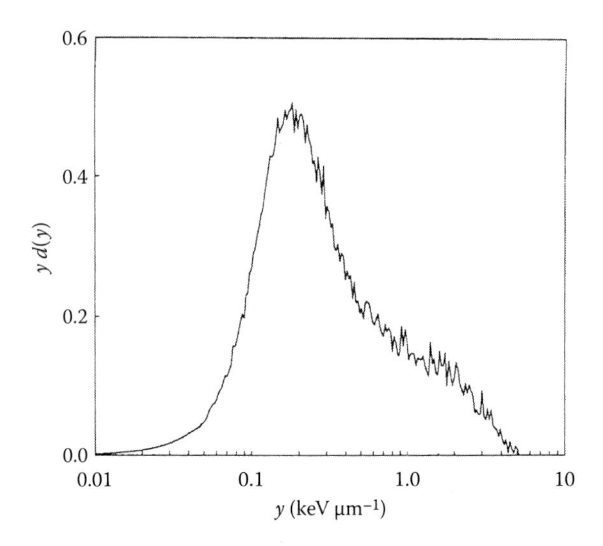

FIGURE 5.17 Single-event dose distribution observed with a wall-less counter inside a steel tank when a 'hot' particle found on the ground after the Chernobyl accident was investigated (Grindborg et al., 1990).

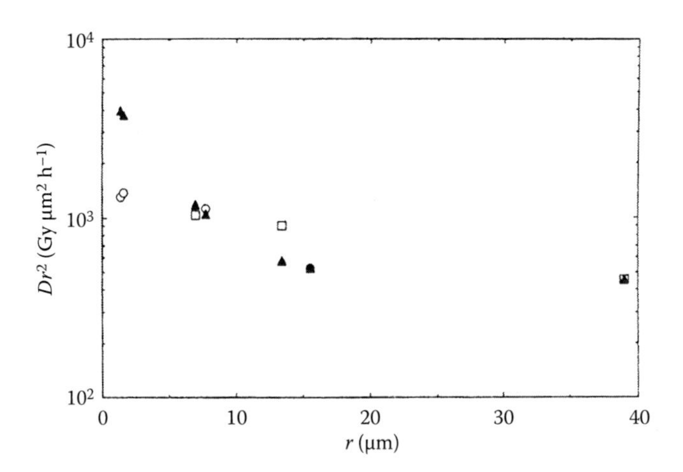

FIGURE 5.18 Determination of the absorbed dose rates at various distances from the 'hot' particle (Figure 5.17). Filled triangles are calculated dose rates (Grindborg et al., 1990). In the figure, the dose rate has been multiplied by the distance from the source squared. For a point source and long-range particles such a plot should yield a flat line. The large deviation from the flat line observed at the shortest distances indicates the presence of low-energy beta particles.

5.9 SUMMARY

In this chapter, we have defined some of the important quantities used in radiation protection. In particular, effective dose and ambient dose equivalent and their interrelationship has been a focus of the discussion. The basic idea that ambient dose equivalent could be used as a surrogate for values of effective dose without significantly underestimating the effective dose was illustrated with situations from the nuclear industry as well as situations on board aircraft. As the effective dose is often used in regulations a measurement of the ambient dose equivalent will in most situations be sufficient for demonstrating compliance.

Measurements of the ambient dose equivalent can be made with TEPCs using both the pulse height analysis (PHA) method and variance-covariance (VC) method. In both methods, simulated tissue volumes between 1 μm and a few micrometres can be used. From the single-event dose distribution of the lineal energy the Q value is derived with the assumption that $y = L$. The mean quality factor is

$$\bar{Q}(y) = \int_{\min}^{\max} Q(y)d(y)dy \tag{5.22}$$

Other relationships can also be used. The absorbed dose in the gas volume is measured simultaneously and the ambient dose equivalent is then equal to the product of the two quantities and a calibration factor. Methods to separate the dose equivalent into a low- and a high-LET dose component have also been described.

In the VC method $\bar{y}_D$ and D are determined. The size of the detector needs to be large enough to ensure multi-event analyses in the radiation field being quantified and a mean quality factor is derived from a given relation to $\bar{y}_D$. The following relationship is usually applicable $Q = a + b\bar{y}_D$. As Q and $\bar{y}_D$ are not proportional to each other, values of a and b have to be chosen with caution. However, with optimised functions agreement between the VC and PHA methods is found to be good. Both methods will allow dose fractions from high- and low-LET components to be determined.

Bibliography

Agosteo S. and Pola A. Silicon microdosimetry. *Radiat. Prot. Dosim.* 143(2–4), 409–415 (2011). In R. Cherubini, F. A. Cucinotta, S. Geradi, H. G. Menzel and P. O'Neill (eds.), *Proceedings of the 15th International Symposium on Microdosimetry*, Verona, Italy, 25–30 October 2009.

Agosteo S., Colautti P., Fanton I., Introini M. V., Moro D., Pola A. and Varoli V. Study of a solid state microdosimeter based on a monolithic silicon telescope: Irradiations with low-energy neutrons and direct comparison with a cylindrical TEPC. *Radiat. Prot. Dosim.* 143(2–4), 432–435 (2011).

Al-Bayati S. N. *The application of experimental microdosimetry to mixed-field neutron-gamma dosimetry*. MASc thesis, University of Ontario Institute of Technology (2012). Available at: http://hdl.handle.net/10155/294.

Alberts W. G., Dietze G., Guldbagge S., Kluge H. and Schuhmacher H. Radiation protection instruments based on tissue equivalent proportional counters: Part II of an international intercomparison. PTB Report PTB-FMRB-117. Braunscweig: Physikalische-Technische Bundesanstalt (1988).

Ali F., Waker A. J. and Waller E. J. Intercomparison of Monte-Carlo radiation transport codes to model TEPC response in low-energy neutron and gamma-ray fields, *Radiat. Prot. Dosim.* 161(1/4), 257–260 (2014).

Amols H. I., Wuu C. S. and Zaider M. On possible limitations of experimental nanodosimetry, *Radiat. Prot. Dosim.* 31(1/4), 125–128 (1990).

Anachkova E., Kellerer A. M. and Roos H. Calibrating and testing tissue equivalent proportional counters with ^{37}Ar. *Radiat. Environ. Biophys.* 33, 353–364 (1994).

Anachkova E., Kellerer A. M. and Roos H. Neutron energy deposition spectra at simulated diameters down to 50 nm. *Radiat. Prot. Dosim.* 70, 207–210 (1997).

Barendsen G. W. Mechanism of action of different ionizing radiation on the proliferative capacity of mammalian cells. In A. Cole (ed.), *Advances in theoretical and experimental biophysics*, Vol. 1 (pp. 167–231). New York: Marcel Dekker (1967).

Barendsen G. W. Responses of cultured cells, tumours and normal tissues to radiations of different Linear Energy Transfer. *Curr. Top. Radiat. Res.* 4, 293–356 (1968).

Barendsen G. W. Mechanisms of cell reproductive death and shapes of radiation dose-survival curves of mammalian cells. *Int. J. Radiat. Biol.* 57(4), 885–896 (1990).

Bartels E. R. and Harder D. The moicrodosimetric regularities of nanometre regions. *Radiat Prot Dosim* 31(1/4), 211–215 (1990).

Bartlett D. T., Drake P., Lindborg L., Klein H., Schmitz Th. and Tichy M. Determination of the neutron and photon dose equivalent at work places in nuclear facilities of Sweden: An SSI-Eurados comparison exercise. Part 2: Evaluation. SSI report 99:13 (1999).

Batterman J. J., Hart G. A. M. and Breuer K. Dose-effect relations for tumour control and complication rate after fast neutron therapy for pelvic tumours. *Br. J. Radiol.* 54, 899–904 (1981).

Beck P., Latocha M., Rollet S., Ferrari A., Pelliccioni M. and Villari R. In L. Lindborg, D. T. Bartlett, P. Beck, I. R. McAulay, K. Schnuer, H. Schraube and F. Spurny (eds.), *Cosmic radiation exposure of aircraft crew: Compilation of measured and calculated data*, Appendix A. Radiation Protection Issue No. 140. European Commission (2004).

Bengtsson L. G. Assessment of dose equivalent from fluctuation of energy depositions. In *Proceedings of the second symposium on microdosimetry.* EUR-4552 (pp. 375–400). Brussels: Commission of the European Communities (1970).

Bengtsson L. G. and Lindborg L. Comparison of pulse height analysis and variance measurement for the determination of dose mean lineal energy. In *Proceedings of the fourth symposium on microdosimetry.* Brussels: Commission of the European Communities (1974).

Benjamin P. W., Kemshall C. D. and Redfearn J. A high resolution spherical proportional counter *Nucl. Instr. Methods* 59, 77 (1968).

Bigildeev E. A. and Lappa A. V. Determination of energy deposited in nanometre sites from ionisation data. *Radiat. Prot. Dosim.* 52, 73–76 (1994).

Boag J.-W. Ionization chambers. In F. H. Attix, W. C. Roesch and E. Tochilin (eds.), *Radiation dosimetry*, Vol. III: Sources, fields, measurements and application (pp. 2–67). New York: Academic Press (1966).

Booz J. and Feinendegen L. E. A microdosimetric understanding of low-dose radiation effects. *Int. J. Radiat. Biol.* 53(1), 13–21 (1988).

Borak T. B., Doke T., Fuse T., Guetersloh S., Heilbronn L., Hara K., Moyers M. et al. Comparisons of LET distributions for protons with energies between 50 and 200 MeV determined using a spherical tissue-equivalent proportional counter (TEPC) and a position-sensitive silicon spectrometer (RRMD-III). *Radiat. Res.* 162, 687–692 (2004).

Braby L. Experimental microdosimetry: History, applications and recent technical advances. *Radiat. Prot. Dosim.* 166(1/4), 3–9 (2015).

Breckow J., Wenning A., Roos H. and Kellerer A. M. The variance-covariance method: Microdosimetry in time-varying low dose-rate radiation fields. *Radiat. Environ. Biophys.* 27, 247–259 (1988).

Brenner D. J. The linear-quadratic model is an appropriate methodology for determining isoeffective doses at large doses per fraction. *Radiat. Oncol.* doi:10.1016/j.semradonc.2008.04.004 (2008).

Brenner D. J. and Sachs R. K. Estimating radiation-induced cancer risks at very low doses: Rationale for using a linear no-threshold approach. *Radiat. Environ. Biophys.* 44, 253–256 (2005).

Brenner D. J. and Ward J. F. Constraints on energy depositions and target size of multiply damaged sites associated with DNA double-strand breaks. *Int. J. Radiat. Biol.* 61, 737–748 (1992).

Brenner D. J. and Zaider M. The application of track calculations to radiology. II. Calculations of microdosimetric quantities. *Radiat. Res.* 98, 14–25 (1984).

Brenner D. J. and Zaider M. Estimating RBEs at clinical doses from microdosimetric spectra. Letter to the Editor. *Med. Phys.* 25(6), 1055–1057 (1998).

Brenner D. J., Doll R., Goodhead D. T., Hall E. J., Land C. E., Little J. B., Lubin J. H. et al. Cancer risks attributed to low doses of ionizing radiation: Assessing what we really know. *PNAS*, 100(24), 13766 (2003).

Bronic I. K. *W*-values in propane-based tissue-equivalent gas. *Radiat. Prot. Dosim.* 70, 33–36, 1997.

Burlin T. E. The characteristics of secondary electron emission and some potential applications to microdosimetry. In *Proceedings of the fourth symposium on microdosimetry*, Verbania EUR5122 (p. 35). Luxembourg: Commission of the European Communities (1974).

Butts J. J. and Katz R. Theory of RBE for heavy ion bombardment of dry enzymes and viruses. *Radiat. Res.* 30, 855–871 (1967).

Campion P. J. The operation of proportional counters at low pressures for microdosimetry. *Phys. Med. Biol.* 16, 611–616 (1971).

Cesari V., Iborra N., De Nardo L., Querini P., Conte V., Colautti P., Tornielli G. and Chauvel P. Microdosimetric measurements of the Nice therapeutic proton beam. In International conference on ocular pathologies therapy with proton beams, Catania, Italy, October 12–13, 2000.

Cesari V., De Nardo L., Colautti P., Tornelli G., Müller-Veggian M. and Donà G. First microdosimetric measurements down to 25 nm LNL report 2000. LNL-INFN(REP)-178/2001, 82–83 (2001).

Chechik R., Breskin A., Shalem C. and Mörmann D. Thick gem-like hole multipliers: Properties and possible applications. *Nucl. Instr. Meth.* A535, 303–308 (2004).

Chechik R., Breskin A. and Shalem C. Thick gem-like multipliers: A simple solution for large area UV-rich detectors. *Nucl. Instr. Meth.* A553, 35–40 (2005).

Chen J. Microdosimetric characteristics of carbon track-segments. *Radiat. Prot. Dosim.* 126(1–4), 445–448 (2007).

Chen J. Microdosimetric characteristics of proton beams from 50 keV to 200 MeV. *Radiat. Prpot. Dosim.* 143(2/4), 436–439 (2011).

Chen J. and Kellerer A. M. Computations to facilitate TEPC calibration with ^{37}Ar. *Nucl. Instr. Methods Phys. Res.* A386, 450–452 (1997).

Chen J., Breckow J., Roos H. and Kellerer A M. Further development of the variance-covariance method. *Radiat. Prot. Dosim.* 31(1/4), 171–174 (1990).

Chiriotti S., Moro D., Conte V., Grosswendt B., Vanhavere F. and Vynckier S. Indirect method to monitor the site size of sealed TEPCs. *Radiat. Meas.* 85, 26–31 (2016).

Clarke R. H. and Valentin J. The history of ICRP and the evolution of its policies *Ann. ICRP* 39, 75–110 (2009).

Crossman J. S. P. and Watt D. E. Inherent calibration of microdosimeters for dose distributions in lineal energy. *Radiat. Prot. Dosim.* 55(4), 295–298 (1994).

d'Errico F., Giusti V., Reginatto M. and Wiegel B. A telescope-design directional neutron spectrometer. *Radiat. Prot. Dosim.* 110(1/4), 533–537 (2004).

De Nardo L., Alkaa A., Khamphan C., Conte V., Colautti P., Ségur P. and Tornelli G. A detector for track-nanodosimetry. *Nucl. Instrum. Methods Phys. Res. A*484, 312–326 (2002).

De Nardo L., Moro D., Colautti P., Conte V., Tornielli G. and Cuttone G. Microdosimetric investigation at the Therapeutic Proton Beam Facility of Catana. *Radiat. Prot. Dosim.* 110(1/4), 681–686 (2004).

De Nardo L., Ferretti A., Colautti P. and Grosseendt B. Bayesian analysis of nanodosimetric ionisation distributions due to alpha particles and protons. *Radiat. Prot. Dosim.* 143(2/4), 459–462 (2011).

Dennis J. A. The application of Jaffe and Lea theories of recombination to the determination of LET and QF with ionization chambers. In H. G. Ebert (ed.), *Proceedings of the first symposium on microdosimetry*, Ispra, Italy, 13–15 November 1967. EUR-3747 (pp. 313–327). Brussels: Commission of the European Communities (1968).

Dietze G., Edwards A. A., Guldbakke S., Kluge H., Leroux J. B., Lindborg L., Menzel H. G., Nguyen V. D., Schmitz Th. and Schuhmacher H. Investigation of radiation protection instrument based on tissue-equivalent proportional counters, Results of an EURADOS-intercomparison. Brussels: Commission of the European Communities (1988).

Drake P., Lindborg L. and Sohlstrand C. Measurement of neutron dose equivalent close to a supposedly stuck transport cask with spent reactor fuel. *Radiat. Prot. Dosim.* 80(4), 411–416 (1998).

Dubeau J. and Waker A. J. Neutron microdosimetric response of a gas electron multiplier. *Radiat. Prot. Dosim.* 128(4), 413–420 (2008).

Edwards A. A., Lloyd D. C. and Prosser J. S. The induction of chromosome aberrations in human lymphocytes by 24 keV neutrons. *Radiat. Prot. Dosim.* 31(1/4), 265–268 (1990).

Eickel R. and Booz J. The influence of counter wall and counter shape on the spectral energy deposition in small volumes by ^{60}Co gamma rays and 200 kV X-rays. *Radiat. Environ. Biophys.* 13, 145–165 (1976).

Eivazi M. T. The application of experimental microdosimetric techniques to the dosimetry of x-rays in the diagnostic energy range, PhD Thesis, The University of Leeds, UK, (1990).

Elsässer T., Weyrather W. K., Friedrich T., Durante M., Iancu G. and Krämer M. Quantification of the relative biological effectiveness for ion-beam therapy: Direct experimental comparison of proton and carbon ion beams and a novel approach for treatment planning. *Int. J. Radiat. Oncol. Biol. Phys.* 88, 103–107 (2010).

Endo S., Onizuka Y., Ishikawa M., Takada M., Sakurai Y., Kobayashi T., Tanaka K., Hoshi M. and Shizuma K. Microdosimetry of neutron field for boron neutroncapture therapy at Kyoto university reactor. *Radiat. Prot. Dosim.* 110 (1/4), 641–664 (2004).

Endo S., Takada M., Tanaka H., Onizuka Y., Tanaka K., Miyahara N., Baba H. et al. Measurement of microdosimetric spectra produced from a 290 MeV/n spread out Bragg peak carbon beam. *Radiat. Environ. Biophys.* 49, 469–475 (2010).

EC (European Commission). Edited by L. Lindborg, D. T. Bartlett, P. Beck, I. R. McAulay, K. Schnuer, H. Schraube and F. Spurny. Radiation protection 140: Cosmic radiation exposure of aircraft crew compilation of measured and calculated data. Final report of EURADOS WG 5 to the Group of Experts established under Article 31 of the Euratom treaty (2004).

EURADOS report. Edited by Th. Schmitz, A. J. Waker, P. Kliauga and H. Zoetelief. Design, construction and use of tissue equivalent proportional counters. *Radiat. Prot. Dosim.* 61(4), 297–404 (1995).

Farahmand M., Bos A. J. J., De Nardo L. and Eijk C. W. E. First microdosimetric measurements with a TEPC based on a GEM. *Radiat. Prot. Dosim.* 110(1/4), 839–843 (2004).

Forsberg B. and Lindborg L. Experimental limitations in microdosimetry measurements using the variance technique. *Radiat. Environ. Biophys.* 19, 125–135 (1981).

Garty G., Schulte R., Shchemelinin S., Leloup C., Assaf G., Breskin A., Chechik R., Bashkitrov V., Milligan J. and Grosswendt B. A nanodosimetric model of radiation-induced clustered damage yields. *Phys. Med. Biol.* 55, 761–781 (2010).

Gerlach R., Roos H. and Kellerer A. M. Heavy ion RBE and microdosimetric spectra. *Radiat. Prot. Dosim.* 99, 413–418 (2002).

Glass W. A. and Gross W. A. Wall-less counters in microdosimetry In *Topics in radiation dosimetry* Suppl. 1. New York: Academic Press (1972).

Goldhagen P., Randers-Pehrson G., Marino S. A. and Kliauga P. Variance–covariance measurements of $\bar{y}_D$ for 15 MeV neutrons in a wide range of site size. *Radiat. Prot. Dosim.* 31(1/4), 167–170 (1990).

Golnik N. Recombination methods in the dosimetry of mixed radiation. Report IAE-20A. Otwock-Świerk, Poland: Institute of Atomic Energy. (1996).

Goodhead D. T. Energy deposition stochastics and track structure: What about the target? *Radiat. Prot. Dos* 55(1/4), 3–15 (2006).

Goodhead D. T. and Nikjoo H. Track structure analysis of ultrasoft X-rays compared to high- and low-LET radiation. *Int. J. Radiat. Biol.* 55, 513–529 (1989).

Goodhead D. T., Thacker J. and Cox R. Effectiveness of 0.3 keV carbon ultrasoft x-rays for the inactivation and mutation of cultured mammalian cells. *Int. J. Radiat. Biol.* 36, 101–114 (1979).

Grindborg J.-E. and Olko P. A comparison of measured and calculated $\bar{y}_D$-values in the nanometre region for photon beams. In D. T. Goodhead, P. O'Neill and H. G. Menzel (eds.), *Microdosimetry: An interdisciplinary approach. Twelfth symposium on microdosimetry* (pp. 387–390). Cambridge, UK: The Royal Society of Chemistry (1997).

Grindborg J. E., Lindborg L., Tilikidis A. and Falk R. Dosimetry around hot particles with microdosimetry techniques. *Radiat. Prot. Dosim.* 31(1/4), 389–394 (1990).

Grindborg J.-E., Samuelson G. and Lindborg L. Variance-covariance measurements in photon beams for simulated nanometre objects, *Radiat. Prot. Dosim.* 61, 193–198 (1995).

Grosswendt B. Formation of ionization clusters in nanometric structures of propane-based tissue-equivalent gas or liquid water by electrons and α-particles. *Radiat. Environ. Biophys.* 41, 103–112 (2002).

Gryzinski M. A., Zielcznski M., and Golnik N. Method for the determination of the ratio of absorbed doses created by different radiations from two sources. *Radiat. Meas.* 45, 1224–1227 (2010).

Gueulette J., Grégoire V., Octave-Prignot M. and Wambersie A. Measurements of radiobiological effectiveness in the 85 MeV proton beam produced at the cyclotron CYCLONE of Louvain-la-Neuve, Belgium. *Radiat. Res.* 145, 70–74 (1996).

Hall E. J., Brenner D. J., Hei T. K. and Miller R. C. The microdosimetric link between oncogenic transformation data with neutrons and charged particles. *Radiat. Prot. Dosim.* 31(1/4), 275–278 (1990).

Hanahan D. and Weinberg R. A. Hallmarks of cancer: The next generation. Leading edge review. *Cell* 144(March 4), 646–674 (2011).

Hawkins R. B. A statistical theory of cell killing by radiation of varying linear energy transfer quality. *Radiat. Res.* 140, 366–374 (1994).

Hawkins R. B. A microdosimetric-kinetic model of cell death from exposure to ionizing radiation of any LET, with experimental and clinical applications. *Int. J. Radiat. Biol.* 69, 739–755 (1996).

Hawkins R. B. A microdosimetric-kinetic theory of the dependence of the RBE for cell death on LET. *Med. Phys.* 25, 1157–1170 (1998).

Hawkins R. B. A microdosimetric-kinetic model for the effect of non-Poisson distribution of lethal lesions on the variation of RBE with LET. *Radiat. Res.* 160, 61–69 (2003).

Hawkins R. B. Mammalian cell killing by ultrasoft x rays and high-energy radiation: An extension of the MK model. *Radiat. Res.* 166, 431–442 (2006).

Hawkins R. B. and Inaniwa T. A microdosimetric-kinetic model for cell killing by protracted continuous irradiation II: Brachytherapy and biological effective dose. *Radiat. Res.* 182, 72–82 (2014).

Hogeweg B. *Microdosimetric measurements and some applications in radiobiology and radiation protection.* PhD thesis, Radiobiological Institute of the Organization for Health Research TNO, Rijswijk, the Netherlands (1978).

Hultqvist M. *Secondary absorbed dose distribution and radiation quality in light ion therapy* PhD thesis, Department of Physics, Stockholm University (2011).

Hultqvist M. and Nikjoo H. *Energy deposition by monoenergetic carbon ions (10–300 MeV/u) in cylindrial targets.* SE-171 76. Stockholm, Sweden: Radiation Biophysics Group, Karolinska Institutet (2010).

Hultqvist M., Lillhök J. E., Lindborg L., Gudowska I. and Nikjoo H. Nanodosimetry in a ^{12}C ion beam using Monte Carlo simulations *Radiat. Meas.* 45, 1238–1241 (2010).

IAEA. *Relative biological effectiveness in ion beam therapy*. International Atomic Agency TRS no. 461, Vienna (2008).

ICRP (International Commission on Radiological Protection). Relative biological effectiveness (RBE), quality factor (Q), and radiation weighting factor (w_R). ICRP Publication 92. *Annals of the ICRP* 33(4) (2003).

ICRP (International Commission on Radiological Protection). The 2007 recommendations of the International Commission on Radiological Protection. Publication 103. *Annals of the ICRP* 37(2–4) (2007).

ICRP (International Commission on Radiological Protection). Conversion coefficients for radiological protection quantities for external radiation exposures. Publication 116. *Annals of the ICRP* 40(2–5) (2010).

ICRP (International Commission on Radiological Protection). Compendium of dose coefficients based on ICRP Publication 60. ICRP Publication 119. *Annals of the ICRP* 41, Suppl. (2012).

ICRU (International Commission on Radiation Units & Measurements). Report 16. Linear energy transfer. Bethesda, MD: ICRU (1970).

ICRU (International Commission on Radiation Units & Measurements). Report 31: Average energy required to produce an ion pair. Bethesda, MD: ICRU (1979).

ICRU (International Commission on Radiation Units & Measurements). Report 36: Microdosimetry. Bethesda, MD: ICRU (1983).

ICRU (International Commission on Radiation Units & Measurements). Report 44: Tissue substitutes in radiation dosimetry and measurement. *J. ICRU* os23(1), (1989).

ICRU (International Commission on Radiation Units & Measurements). Report 51: Quantities and units in radiation protection dosimetry. Bethesda, MD: ICRU (1993).

ICRU (International Commission on Radiation Units & Measurements). Report 57: Conversion coefficients for use in radiological protection against external radiation. Bethesda, MD: ICRU (1998).

ICRU (International Commission on Radiation Units & Measurements). Report 84: Reference data for the validation of doses from cosmic-radiation exposure of aircraft crew. *J. ICRU* 10(2), 1–35 (2010).

ICRU (International Commission on Radiation Units & Measurements). Report 85a revised: Fundamental quantities and units for ionizing radiation. *J. ICRU* 11(1) (2011).

ICRU (International Commission on Radiation Units & Measurements). Report 86: Quantification and reporting of low-dose and other heterogeneous exposures. *J. ICRU* 11(2), 1–77 (2011).

IMBA®. Professional Plus Internal Dosimetry Software, Public Health England. Available at: https://www.pheprotectionservices.org.uk/imba/ (Accessed February 2017).

ISO (International Organization for Standardization). 4037-3:1999: X and gamma reference radiation for calibrating dosemeters and doserate meters and for determining their response as a function of photon energy – Part 3:

Calibration of area and personal dosemeters and the measurement of their response as a function of energy and angle of incidence. Geneva, Switzerland: ISO (1999).

Jacobi W. The concept of the effective dose: A proposal for the combination of organ doses. *Radiat. Environm. Biophys.* 12, 101–109 (1975).

JCGM (Joint Committee for Guides in Metrology). Evaluation of measurement data. *Guide to the expression of uncertainty in measurement JCGM*, available on the website of Bureau International des Poids et Measures www.bipm.org.

Kase Y., Kanai T., Matsumoto Y., Furusawa Y., Okamoto H., Asaba T., Sakama M. and Shinoda H. Microdosimetric measurements and estimation of human cell survival for heavy-ion beams. *Radiat. Res.* 166, 629–638 (2006).

Katz R., Sharma S. C. and Hamayoonfar M. Cellular inactivation by heavy ions, neutrons, and pions. In H. G. Ebert (ed.), *Third symposium on microdosimetry*, Stresa Italy, 18–22 October, 1971 (pp. 267–287). Luxembourg: Commission of the European Communities, Directorate General for Dissemination of Information, Centre for Information and Documentation (1972).

Kellerer A. M. *Mikrodosimetrie.* GSF-Bericht B-1 (Forschungszentrum fur Umwelt und Gesundheit) Neuherberg/München, Germany: Gesellschaft für Strahlen und Umweltforschung (1968).

Kellerer A. M. Event simultaneity in cavities: Theory of the distortions of energy deposition in proportional counters. *Radiat. Res.* 48, 216–233 (1971a).

Kellerer A. M. An assessment of the wall effect in microdosimetric measurements. *Radiat. Res.* 47, 377–386 (1971b).

Kellerer A. M. Chord-length distributions and related quantities for spheroids. *Radiat. Res.* 98, 425–437 (1984).

Kellerer A. M. Fundamentals of microdosimetry. In Kr. Kase, B. E. Bjärngard and F. H. Attix (eds.), *The dosimetry of ionizing radiation*, Vol. 1 (pp. 78–158). Orlando, FL: Academic Press (1985).

Kellerer A. M. Generalization of the variance-covariance method for microdosimetric measurements. I. Basic equations and estimation formulae for constant and time-varying fields. *Radiat. Environ. Biophys.* 35, 111–115 (1996a).

Kellerer A. M. Generalization of the variance-covariance method for microdosimetric measurements. II. Formulae for varying dose-rate ratio in the detector and synopsis of results. *Radiat. Environ. Biophys.* 35, 117–119 (1996b).

Kellerer A. M. Generalization of the variance-covariance method for microdosimetric measurements. III. Formulae for varying dose-rate ratio in the detector and synopsis of results. *Radiat. Environ. Biophys.* 35, 145–152 (1996c).

Kellerer A. M. Microdosimetry: Reflections on Harald Rossi. *Radiat. Prot. Dos* 99 (1/4), 17–22 (2002).

Kellerer A. M. and Rossi H. H. The theory of dual radiation action. *Curr. Top. Radiat. Res. Q.* Ebert M. and Howard A. (eds.), 8, 85–158 (1972).

Kellerer A. M. and Rossi H. H. On the determination of microdosimetric parameters in time varying radiation fields: The variance-covariance method. *Radiat. Res.* 97, 237–245 (1984).

Kellerer A. M., Lam Y.-M. P. and Rossi H. H. Biophysical studies with spatially correlate ions 4. Analysis of cell survival data for diatomic deuterium. *Radiat. Res.* 83, 511–528 (1980).

Kliauga P. Microdosimetry at middle age: Some old experimental problems and new aspirations. *Radiat. Res.* 124, S5–S15 (1990a).

Kliauga P. Measurement of single event energy deposition spectra at 5 nm to 250 nm simulated site sizes. *Radiat. Prot. Dosim.* 31(1/4), 119–123 (1990b).

Kliauga P., Amols H. and Lindborg L. Microdosimetry of pulsed radiation fields employing the variance method. *Radiat. Res.* 105, 129–137 (1986).

Knoll G. F. *Radiation detection and measurement*, 4th edition. New York: Wiley (2010).

Konovalov V. V., Raitsimring A. M. and Tsvetkov, Yu. D. Thermalization lengths of "subexcitation electrons" in water determined by photoinjection from metals into electrolyte solutions. *Radiat. Phys. Chem.* 32, 623–632 (1988).

Kunz A., Arend E., Dietz E., Gerdung S., Grillmaier R. E., Lim T. and Pihet P. The Homburg area neutron dosmeter HANDI: Characteristics and optimisation of the operational instrument. *Radiat. Prot. Dosim.* 44(1/4), 213–218 (1992).

Kyllönen J. and Lindborg L. Photon and neutron dose discrimination using low pressure proportional counters with graphite and A150 walls. *Radiat. Prot. Dosim.* 125(1/4), 314–317 (2007).

Kyllönen J.-E., Lindborg L. and Samuelsson G. The response of the Sievert instrument in neutron beams up to 180 MeV. *Radiat. Prot. Dosim.* 94(3), 227–232 (2001).

Laczkó G. *Investigation of the radial ionization distribution of heavy ions with an optical particle track chamber and Monte-Carlo simulations.* Dissertation, Frankfurt, Germany: J W Goethe-Universität (2006).

Liamsuwan T., Uehara S. and Nikjoo H. *Energy deposition by monoenergetic protons (0.3–300 MeV) in cylindrical targets.* Monograph 2010/2. Available from Radiation Biophysics Group, Department of Oncology-Pathology, Karolinska Institutet, Stockholm, Sweden (2010).

Liamsuwan T., Hultqvist M., Lindborg L., Uehara S. and Nikjoo H. Microdosimetry of protons and carbon ions. *Med. Phys.* 41, 081721 (2014).

Lillhök J. E. *The microdosimetric variance-covariance method used for beam quality characterization in radiation protection and radiation therapy.* Thesis, Medical Radiation Physics, Stockholm University and Karolinska Institutet (2007a).

Lillhök J. E., Grindborg J.-E., Lindborg L., Gudowska I., Alm Carlsson G., Söderberg J., Kopec M. and Medin J. Nanodosimetry in a clinical neutron therapy beam using the variance-covariance method and Monte Carlo simulations. *Phys. Med. Biol.* 52, 4953–4966 (2007b).

Lillhök J., Beck P., Bottolier F., Latocha M., Lindborg L., Roos H., Roth J. et al. A comparison of ambient dose equivalent meters and dose calculations at constant flight conditions. *Radiat. Meas.* 42, 323–333 (2007c).

Lindborg L. Microdosimetry in high energy electron and ^{60}Co gamma ray beams for radiation therapy. In *Proceedings of the fourth symposium on microdosimetry*, J. Booz, H. G. Ebert, R. Eickel and A. Waker (eds.), Verbania Pallanza, Italy, 24–28 September 1973. EUR 5122 d-e-f (pp. 799–819). Luxembourg: European Commission (1974).

Lindborg L. Microdosimetry measurements in beams of high energy photons and electrons: Technique and results. In *Fifth symposium on microdosimetry*, J. Booz, H. G. Ebert and B. G. R. Smith (eds.), Verbania Pallanza, Italy, 22–26 September 1975. EUR 5452 d-e-f (pp. 347–376). Luxembourg: European Commission (1976).

Lindborg L. and Brahme A. Influence of microdosimetric quantities on observed dose-response relationships in radiation therapy. *Radiat. Res.* 124, S23–S28 (1990).

Lindborg, L. and Grindborg J. E. Nanodosimetric results and radiotherapy beams: A clinical application? *Radiat. Prot. Dosim.* 70(1/4), 541–546 (1997).

Lindborg L. and Nikjoo H. Microdosimetry and radiation quality determinations in radiation protection and radiation therapy. *Radiat. Prot. Dosim.* 143, 402–408 (2011).

Lindborg L., Kliauga P., Marino S. and Rossi H. H. Variance-covariance measurements of the dose mean lineal energy in a neutron beam. *Radiat. Prot. Dosim.* 13(1/4), 347–351 (1985).

Lindborg L., Marino S., Kliauga P. and Rossi H. H. Microdosimetric measurements and the variance-covariance method: Some experimental experience. *Radiat. Environ. Biophys.* 28, 251–263 (1989).

Lindborg L., Bartlett D., Drake P., Klein H., Schmitz Th. and Tichy M. Determination of neutron and photon dose equivalent at workplaces in nuclear facilities in Sweden: A joint SSI_EURADOS comparison exercise. *Radiat. Prot. Dosim.* 61(1/3), 89–100 (1995).

Lindborg L., Bolognese-Milsztajn T., Boschung M., Coeck M., Curzio G., d'Errico F., Fiechtner A. et al. Application of workplace correction factors to dosemeter results for the assessment of personal doses at nuclear facilities. *Radiat. Prot. Dosim. 124*(3), 213–218 (2007).

Lindborg L., Hultqvist M., Carlsson Tedgren Å. and Nikjoo H. Lineal energy and radiation quality in radiation therapy: Model calculations and comparison with experiment. *Phys. Med. Biol.* 58, 3089–3105 (2013).

Lindell B. *Pandoras ask*. Stockholm, Sweden: Atlantis AB, ISBN 91-7486-347-9 (1996).

Lindell B. *Geschichte der Strahlenforschung, T. 1, Pandoras Büchse: die Zeit vor dem Zweiten Weltkrieg*. Bremen, Germany: Aschenbeck und Isensee, ISBN 3-89995-082-8 (2004).

Loncol T., Cosgrove V., Denis J. M., Gueulette J., Mazal A., Menzel H. G., Pihet P. and Sabattier R. Radiobiological effectiveness of radiation beams with broad LET spectra: Microdosimetric analysis using biological weighting functions. *Radiat. Prot. Dosim.* 52(1–4), 347–352 (1994).

Luszik-Bhadra M., Reginatto M. and Lacoste V. Measurement of energy and direction distribution of neutron and photon fluences in workplace fields. *Radiat. Prot. Dosim.* 110 (1/4), 237–241 (2004).

Magrin G., Colautti P. and Tornielli G. *Variance-covariance technique for monitoring the TOP proton beam quality.* INFN Laboratori Nazionali di Legnaro L.N.L.-I.N.F.N. (REP) 156/2000, Legnaro, Italy (2000).

Makrigiorgos M. and Waker A. J. Measurement of the restricted LET of photon sources (5 keV–1.2 MeV) by the recombination method: Theory and practice. *Phys. Med. Biol.*, 36(5), 543–554 (1986).

Matsufuji N., Kanai T., Kanematsu N., Miyamoto T., Baba T., Kamada T., Kato H., Yamada S., Mizoe J. and Tsujii H. Specification of carbon ion dose at the National Institute of Radiological Sciences (NIRS). *J. Radiat. Res.* 48, A81–86 (2007).

Mayer S., Golnik N., Kyllönen J. E., Menzel H. G. and Otto Th. Dose equivalent measurements in a strongly pulsed high-energy radiation field. *Radiat. Prot. Dosim.* 110(1–4), 759–762 (2004).

Meesungnoen J., Jay-Gerin J. P., Filali-Mouhim A. and Mankhetkorn S. Low-energy electron penetration range in liquid water. *Radiat. Res.* 158, 657–660 (2002).

Melinder A. *Gas gain properties and temperature dependence of a tissue equivalent proportional counter filled with methane and propane based tissue equivalent gas.* Thesis for master of science in medical radiation physics, Department of Medical Physics, Karolinska Institutet, Stockholm University (1999).

Menzel H. G. *Anwendungsmöglichkeiten der experimentellen mikrodosimetrie für untersuchungen zum problem der strahlenqualität bei der tumortherapie mit schnellen neutronen.* PhD thesis, Universität des Saarlandes, Saarbrücken, 1981.

Menzel H. G., Bühler G. and Schumacher H. Investigation of basic uncertainties in the experimental determination of microdosimetric data. In J. Booz and H. G. Ebert (eds.), *Proceedings of the eighth symposium on microdosimetry.* EUR-8395 (pp. 1061–1072). Jülich: Commission of the European Communities (1982).

Menzel H. G., Lindborg L., Schmitz Th., Schuhmacher H. and Waker A. J. Intercomparison of dose equivalent meters based on microdosimetric techniques: Detailed analysis and conclusions. *Radiat. Prot. Dosim.* 29(1/2), 55–68 (1989).

Menzel H. G., Pihet P. and Wambersie A. Microdosimetric specification of radiation quality in neutron radiation therapy. *Int. J. Radiat. Oncol. Biol. Phys.* 57, 865–883 (1990).

Michaud M., Wen A. and Sanche L. Cross sections for low-energy (1–100 eV) electron elastic and inelastic scattering in amorphous ice. *Radiat. Res.* 159, 3–22 (2003).

Mitaroff A. and Silari M. The CERN-EU high-energy reference field (CERF) facility for dosimetry at commercial flight altitudes and in space. *Radiat. Prot. Dosim.* 102, 7–22 (2002).

Moro D., Colautti P., Lollo M., Esosito J., Conte V., De Nardo L., Ferreti A. and Ceballos C. BNCT Dosimetry performed with a mini twin tissue-equivalent proportional counter (TEPC) *Appl. Radiat. Isotop.* 67, S171–S174 (2009).

Moro D., Chiriotti S., Conte V., Colautti P. and Grosswendt B. Lineal energy calibration of a spherical TEPC. *Radiat. Prot. Dosim.* 166(1/4), 233–237 (2015).

NASA. Edited by J. W. Wilson, W. Jones, D. L. Maiden and P. Goldhagen. *Atmospheric ionizing radiation (AIR): Analysis, results, and lessons learned from the June 1997 ER-2 campaign* NASA/CP-2003–212155. Hanover, MD: NASA Center for AeroSpace Information (CASI) (February 2003).

Natrella M. G. *Experimental statistics.* NBS Handbook 91. Washington, DC: National Bureau of Standards (1966).

Navarro F. V. *Micro/nanometric scale study of energy deposition and its impact on the biological response for ionizing radiation, Brachytherapy radionuclides proton and carbon ion beams*, PhD thesis, Faculty of Medicine 1188, Acta Universitatis Upsaliensis, Uppsala. ISSN 1651-6206, ISBN 978-91-554-9495-7 (2016).

NCRP Report 121. Principles and application of collective dose in radiation protection. Bethesda, MD: National Council on Radiation Protection and Measurements (1995).

Nikjoo H. and Liamsuwan T. Biophysical basis of ionizing radiation. In *Comprehensive biomedical physics, 9. Radiation therapy: Physics and treatment optimization* (pp. 65–104). A. Brahme (ed.). Amsterdam: Elsevier (2014).

Nikjoo H., Goodhead D., Charlton D. E. and Paretzke H. *Energy deposition by monoenergetic electrons in cylindrical volumes.* Harwell: MRC Radiobiology Unit (1994).

Nikjoo H., Uehara S., Pinsky L. and Cucinotta Francis A. Modelling and calculations of the response of tissue equivalent proportional counter to charged particles. *Radiat. Prot. Dos.* 126(1/4), 512–518 (2007).

Nikjoo H. and Lindborg L. RBE of low energy electrons and photons: Topical review. *Phys. Med. Biol.* 55, R65–R109 (2010).

Nikjoo H., Khvostunov I. K. and Cucinotta F. A. The response of tissue equivalent proportional counters to heavy ions. *Radiat. Res.* 157, 435–445 (2002).

Nikjoo H., Uehara S., Emfietzoglou D. and Pinsky L. A database of frequency distributions of energy depositions in small-size targets by electrons and ions. *Radiat. Prot. Dosim.* 143(2/4), 145–151 (2011).

Nikjoo H., Uehara S. and Emfietzoglou D. *Interaction of radiation with matter.* Boca Raton, FL: CRC Press (2012).

Orchard G. M., Chin K., Prestwich, W. V., Waker A. J. and Byun S. H. Development of a thick gas electron multiplier for microdosimetry. *Nucl. Instr. Methods A* 638(1), 35, 122–126 (2011).

Palm S., Humm J. L., Rundqvist R. and Jacobsson L. Microdosimetry of astatine-211 single-cell irradiation: Role of daughter polonium-211 diffusion. *Med. Phys.* 31(2), 218–225 (2004).

Palmans H., Rabus H., Belchior A. L., Bug M. U., Galer S., Giesen U., Gonon G., Gruel G. et al. Future development of biologically relevant dosimetry. *Br. J. Radiol.* 88(1045), 20140392 (2015).

Perez-Nunez D. and Braby L. A. Replacement tissue-equivalent proportional counter for the international space station. *Radiat. Prot. Dosim.* 143(2/4), 394–397 (2011).

Periale L., Preskov V., Carlso P., Francke P., Pavlopoulos P. and Pietropaolo F. Detection of the primary scintillation light from dense Ar, Kr,and Xe with novel photosensitive gaseous detectors. *Nucl. Instr. Methods A* 478, 377–383 (2002).

Pihet P. *Étude microdosimétrique des faisceaux de neutrons de haute énergie. Applications dosimétriques et radiobiologiques.* PhD thèse, Université Catholique de Louvain, Louvain-la-Neuve (1989).

Pihet P. and Menzel H. G. Response to the Letter to the Editor. Estimating RBEs at clinical doses from microdosimetric spectra [*Med. Phys.* 25, 1055 (1998)]. *Med. Phys.* 26(5), 848–852 (1999).

Pihet P., Menzel H. G., Schmidt R., Beauduin M. and Wambersie A. Biological weighting function for RBE specification of neutron therapy beams. Intercomparison of 9 European Centres. In *Proceedings of the tenth symposium on microdosimetry*, Rome, Italy, 21–26 May 1989. *Radiat Prot. Dosim.* 31(1/4) (1990).

Prokopovich D. A., Reinhard M. I., Cornelius I. M. and Rosenfeld A. B. SOI microdosimetry for mixed-field radiation protection. *Radiat. Meas.* 43, 1054–1058 (2008).

Pszona S., Kula J., and Marjanska S. A new method for measuring ion clusters produced by charged particles in nanometre track sections of DNA size. *Nucl Instrum Methods Phys Res A* 447, 601–660 (2000).

Rademacher S. E., Borak T. B., Zeitlin C., Heilbronn L. and Miller J. Wall effects observed in tissue-equivalent proportional counters from 1.05 GeV/nucleon Iron-56 particles. *Radiat. Res.* 149, 387–395 (1998).

Reginatto M., Tanner R. and Vanhavere F. *Evaluation of individual dosimetry in mixed neutron and photon radiation fields.* PTB Berichte PTB-N-49. Braunschweig, Germany (2006).

Roesch W. C. Microdosimetry of internal sources. *Radiat. Res.* 70, 494–510 (1977).

Rollet S., Angelone M., Magrin G., Marinelli M., Milani E., Pilon M., Prestopino G., Verona C. and Verona-Rinati G. A novel microdosimeter based on artificial single crystal diamond. *IEEE Trans. Nucl. Sci.* 59(5), 2049 (2012).

Rosenfeld A. B. Novel detectors for silicon based microdosimetry, their concepts and applications. *Nucl. Instr. Methods Phys. Res. A* 809, 156–170 (2016).

Rossi H. H. Specification of radiation quality. *Radiat. Res.* 10, 522–531 (1959).

Rossi H. H. and Rosenzweig W. A device for the measurement of dose as a function of specific ionization. *Radiology* 64(3), 404–411 (1955a).

Rossi H. H. and Rosenzweig W. Measurements of neutron dose as a function of linear energy transfer. *Radiat. Res.* 2(5), 417–425 (1955b).

Rossi H. H. and Zaider M. *Microdosimetry and its applications.* Berlin and New York: Springer-Verlag (1996).

Rossi H. H., Biavati M. H. and Gross W. Local energy density in irradiated tissues. 1. Radiobiological significance. *Radiat. Res.* 15(4), 431–439 (1961).

Sabattier R. Radiobiological effectiveness of radiation's beams with broad LET spectra: Microdosimetric analysis using biological weighting functions. *Radiat. Prot. Dosim.* 52(1/4), 347–352 (1994).

Santa Cruz G. A., Palmer M. R., Matatagui E. and Zamenhof R. G. A theoretical model for event statistics in microdosimetry. I: Uniform distribution of heavy ion tracks. *Med. Phys.* 28, 988–996 (2001a).

Santa Cruz G. A., Palmer M. R., Matatagui E. and Zamenhof R. G. A theoretical model for event statistics in microdosimetry. II: Nonuniform distribution of heavy ion tracks. *Med Phys.* 28, 997–1005 (2001b).

Sato T., Watanabe R., Kase Y., Tsuruoka C., Suzuki M., Furusawa Y. and Niita K. Analysis of cell-survival fractions for heavy-ion irradiations based on microdosimetric kinetic model implemented in the particle and heavy ion transport code system. *Radiat Prot. Dosim.* 143(2/4), 491–496 (2011).

Sauli F. GEM: A new concept for electron amplification in gas detectors. *Nucl. Instrum. Methods A* 386, 531–534 (1997).

Sauli F. Development and applications of gas electron multiplier detectors. *Nucl. Instrum. Methods A* 505, 195–198 (2003).

Sauli F. Recent topics in gaseous detectors. *Nucl. Instrum. Methods A* 623, 29–34 (2010).

Scholz M. and Kraft G. Calculation of heavy ion inactivation probabilities based on track structure, x-ray sensitivity and target size. *Radiat. Prot. Dosim.* 52 (1/4), 29–33 (1994).

Scholz M., Kellerer A. M., Kraft-Weyrather W. and Kraft G. Computation of cell survival in heavy ion beams for therapy: The model and its approximation. *Radiat. Environ. Biophys.* 36, 59–66 (1997).

Schrewe U. J., Newhauser W. D., Brede H. J. and DeLuca P. M. Experimental kerma coefficients and dose distributions of C, N, O, Mg, Al, Si, Fe, Zr, A-150 plastic, Al_2O_3, AlN, SiO_2 and ZrO_2 for neutron energies up to 66 MeV. *Phys. Med. Biol.* 45, 651–683 (2000).

Schuhmacher H. and Dangendorf V. Experimental tools for track structure investigations: New approaches for dosimetry and microdosimetry. *Radiat. Prot. Dosim.* 99(1/4), 317–323 (2002).

Schuhmacher H., Menzel H. G. and Blattmann H. Energy deposition of negative pions and their application to radiation therapy. *Radiat. Environm. Biophys.* 16, 239–244 (1979).

Schuhmacher H., Bartlett D., Bolognese-Milsztajn T., Boschung M., Coeck M., Curzio G., d'Errico F. et al. Experimental basis for optimisation of the wall thickness of microdosimetric counters in radiation protection. *Radiat. Prot. Dosim.* 13(1/4), 341–345 (1985).

Schuhmacher H., Bartlett D., Bolognese-Milsztajn T., Boschung M., Coeck M., Curzio G., d'Errico F. et al. *Evaluation of individual dosimetry in mixed neutron and photon radiation fields.* PTB, Braunschweig, Germany, Berichte PTB-N-49 ISSN 0936-0492 ISBN 3-86509-503-8 (2006).

Scott M. C., de Aro A., Green S. and Taylor G. C. Elemental synthesis of real tissue microdosimetric responses to high energy neutrons: Principles and limitations. In *Proceedings of the medical satellite meeting of the second European particle accelerator conference EPAC 90*. P. Chauvel, A. Wambersie and P. Mandrillon (eds.), Nice, France, 14–16 June 1990 (pp. S23–S25). (1990). Available at: http://epaper.kek.jp/e90/PDF/SPECIAL.PDF.

Ségur P., Olko P. and Colautti P. Numerical modelling of tissue equivalent proportional counters. *Radiat. Prot. Dosim.* 61(4), 323–350 (1995).

Shchemelinin S., Breskin A., Chechik R., Colautti P., Schulte R. W. M. First measurements of ionization clusters on the DNA scale in a wall-less sensitive volume. *Radiat. Prot. Dosim.* 82, 43–50 (1999).

Shonka R. F., Rose J. E. and Failla G. Conducting plastic equivalent to tissue, air and polystyrene. In Second United Nations International Conference on Peaceful Uses of Atomic Energy, 21, 184. New York: United Nations (1958).

Sievert R. Untersuchungen über die an verschiedenen Schwedischen Krankenhäusern zur Erreichung des Hauterythems gebräuchlichen Röntgenstrahlenmengen, unter Einführung der "R"-Einheit. *Acta Radiol.* 7, 461–473 (1926).

Smathers J. B., Otte V. A., Smith A. R., Almond P. R., Attix F. H., Spokas J. J., Quam W. M. and Goodman L. J. Composition of A-150 tissue-equivalent plastic. *Med. Phys.* 4 (1), 74–77 (1977).

Solevi P., Magrin G., Moro D. and Mayer R. Monte-Carlo study of microdosimetric diamond detectors. *Phys. Med. Biol.* 60, 7069–7083 (2015).

Sullivan A. H. The estimation of the local energy loss distribution of ionizing radiation from observations of ion recombination in gas. In *Proceedings of the first symposium on microdosimetry*, Ebert H. G. (eds.), 1990 Ispra, Italy, 13–15 November 1967. EUR 3747 (pp. 295–312). Brussels: Commission of the European Communities (1968).

Tilikidis A., Lind B., Nafstadius P. and Brahme A. An estimation of the relative biological effectiveness of 50 MV bremsstrahlung beams by microdosimetric techniques. *Phys. Med. Biol.* 41, 55–69 (1996).

Uehara S. and Nikjoo H. Monte Carlo simulation of water radiolysis for low-energy charged particles. *J. Radiat. Res.* 47, 69–71 (2006).

Varma M. N. Calibration of proportional counters in microdosimetry. In J. Booz and H. G. Ebert (eds.), *Proceedings of the eighth symposium on microdosimetry* EUR 8395 (pp. 1051–1059). Luxembourg: Commission of the European Communities (1983).

Verma P. K. and Waker A. J. Optimization of the electric field distribution in a large-volume tissue equivalent proportional counter. *Phys. Med. Biol.* 37(10), 1837–1846 (1992).

Villegas F. and Ahnesjö A. Reply to the comment on Monte Carlo calculated microdosimetric spread for cell nucleus-sized targets exposed to brachytherapy ^{125}I and ^{192}Ir sources and ^{60}Co cell irradiation. *Phys. Med. Biol.* 61(13), 5103–5106 (2016).

Villegas F., Tilly N. and Ahnesjö A. Monte Carlo calculated microdosimetric spread for cell nucleus-sized targets exposed to brachytherapy ^{125}I and ^{192}Ir sources and ^{60}Co cell irradiation *Phys. Med. Biol.* 58, 6149–6162 (2013).

von Engel A. *Ionized gases.* New York: American Vacuum Society Classics. Reprinted from 1955 original by American Institute of Physics (1994).

Wagner R. S., Grosswendt B., Harvey I. R., Mill A. J., Selbach H.-J. and Siebert B. R. L. Unified conversion functions for the new ICRU operational radiation quantities. *Radiat. Prot. Dosim.* 12, 231–235 (1985).

Waker A. J. Gas gain characteristics of some walled proportional counters used in microdosimetry. In J. Booz and H. G. Ebert (eds.), *Proceedings of the eighth symposium on microdosimetry. EUR 8395*, 27 September–1 October 1982, Jülich, Germany (pp. 1017–1030). Luxembourg: Commission of the European Communities (1982).

Waker A. J. Experimental uncertainties in microdosimetric measurements and an examination of the performance of three commercially produced proportional counters. *Nucl. Instr. Method. A* 234, 354–360 (1985).

Waker A. J. An investigation of the characteristics of a spherical single-wire proportional counter used for experimental microdosimetry. *Nucl. Instr. Methods A* 243, 561–566 (1986).

Waker A. J. Some electrical and fabrication properties of virgin and recycled A-150 tissue equivalent plastic. *Phys. Med. Biol.* 33(1), 157–164 (1988).

Waker A. J., Aslam and Lori J. Design of a multi-element TEPC for neutron monitoring, *Radiat. Prot. Dosim.* 143(2/4), 463–466 (2011).

Waker A. J., Mahilrajan T. and Sandhu H. Environmental microdosimetry: Microdosimetric characterisation of low dose exposures. *Radiat. Prot. Dosim.* 166(1/4), 204–209 (2015).

Waligórski M. P. R., Grzanka L. and Korcyl M. The principles of Katz's cellular track structure radiobiological model. *Radiat. Prot. Dosim.* 166(1/4), 49–55 (2015).

Walstam R. *Hänt - Men kanske mindre känt - om strålbehandling.* Swedish Association for Medical Physics, distributed through Amersham Health AB (2002).

Wambersie A., Menzel H. G., Gahbauer R. A., Jones D. T., L., Michael B. D. and Paretzke H. Biological weighting of absorbed dose in radiation therapy. *Radiat. Prot. Dosim.* 99, 445–452 (2002).

Watt D. E. *Quantities for dosimetry of ionizing radiations in liquid water.* London: Taylor & Francis (1996).

Zaider M. and Brenner D. J. The application of track calculations to radiobiology. III Analysis of the molecular beam experiment results. *Radiat. Res.* 100, 245–256 (1984).

Zaider M. and Rossi H. H. Definitions of physical and biological low dose. *Int. J. Radiat. Biol.* 74(5), 633–637 (1998).

Zaider M., Brenner D. J., Hanson K. and Minerbo G. N. An algorithm for determining the proximity distribution from dose-average lineal energies. *Radiat. Res.* 91, 95–103 (1982).

Zielczynski M. Use of columnar recombination for the determination of relative biological efficiency of radiation. *Nukleonika* 7, 175–182 (1962).

Index

D

E

Q

R

S

T

V

W

Z